P9-AFP-839

Hilbert's Tenth Problem

Foundations of Computing

Michael Garey and Albert Meyer, editors

Complexity Issues in VLSI: Optimal Layouts for the Shuffle-Exchange Graph and Other Networks, Frank Thomson Leighton, 1983

Equational Logic as a Programming Language, Michael J. O'Donnell, 1985

General Theory of Deductive Systems and Its Applications, S. Yu Maslov, 1987

Resource Allocation Problems: Algorithmic Approaches, Toshihide Ibaraki and Naoki Katoh, 1988

Algebraic Theory of Processes, Matthew Hennessy, 1988

PX: A Computational Logic, Susumu Hayashi and Hiroshi Nakano, 1989

The Stable Marriage Problem: Structure and Algorithms, Dan Gusfield and Robert Irving, 1989

Realistic Compiler Generation, Peter Lee, 1989

Single-Layer Wire Routing and Compaction, F. Miller Maley, 1990

Basic Category Theory for Computer Scientists, Benjamin C. Pierce, 1991

Categories, Types, and Structures: An Introduction to Category Theory for the Working Computer Scientist, Andrea Asperti and Giuseppe Longo, 1991

Semantics of Programming Languages: Structures and Techniques, Carl A. Gunter, 1992

The Formal Semantics of Programming Languages: An Introduction, Glynn Winskel, 1993

Exploring Interior-Point Linear Programming: Algorithms and Software, Ami Arbel, 1993

Theoretical Aspects of Object-Oriented Programming: Types, Semantics, and Language Design, edited by Carl A. Gunter and John C. Mitchell, 1993

Hilbert's Tenth Problem, Yuri V. Matiyasevich, 1993

Hilbert's Tenth Problem

Yuri V. Matiyasevich

with a foreword by Martin Davis

The MIT Press
Cambridge, Massachusetts
London, England

This book was printed and bound in the United States of America.

Library of Congress Cataloging-in-Publication Data

Matiiasevich, IU. V. (IUrii V.)
[Desiataia problema Gil'berta. English]
Hilbert's tenth problem / Yuri V. Matiyasevich ; with a foreword by Martin Davis.
p. cm. — (Foundations of computing)
Includes bibliographical references and indexes.
ISBN 0-262-13295-8
1. Hilbert's tenth problem. 2. Computable functions. I. Title. II. Series.
QA242.M399 1993
512'.7—dc20 93-28107
CIP

Contents

Series Foreword ix

A Note on the Translation xi

Foreword xiii

Preface to the English Edition xviii

Preface xix

1 Principal Definitions 1

1.1 Diophantine equations as a decision problem 1

1.2 Systems of Diophantine equations 2

1.3 Solutions in natural numbers 4

1.4 Families of Diophantine equations 6

1.5 Logical terminology 9

1.6 Some simple examples of Diophantine sets, properties, relations, and functions 12

2 Exponentiation Is Diophantine 19

2.1 Special second-order recurrent sequences 19

2.2 The special recurrent sequences are Diophantine (basic ideas) 21

2.3 The special recurrent sequences are Diophantine (proof) 26

2.4 Exponentiation is Diophantine 31

2.5 Exponential Diophantine equations 33

3 Diophantine Coding 41

3.1 Cantor numbering 41

3.2 Gödel coding 42

3.3 Positional coding 44

3.4 Binomial coefficients, the factorial, and the prime numbers are Diophantine 45

3.5 Comparison of tuples 47

3.6 Extensions of functions to tuples 49

4 Universal Diophantine Equations 57
4.1 Basic definitions 57
4.2 Coding equations 59
4.3 Coding possible solutions 61
4.4 Computing the values of polynomials 62
4.5 Universal Diophantine equations 64
4.6 Diophantine sets with non-Diophantine complements 65

5 Hilbert's Tenth Problem Is Unsolvable 71
5.1 Turing machines 71
5.2 Composition of machines 73
5.3 Basis machines 75
5.4 Turing machines can recognize Diophantine sets 83
5.5 Diophantine simulation of Turing machines 85
5.6 Hilbert's Tenth Problem is undecidable by Turing machines 92
5.7 Church's Thesis 94

6 Bounded Universal Quantifiers 103
6.1 First construction: Turing machines 103
6.2 Second construction: Gödel coding 104
6.3 Third construction: summation 109
6.4 Connections between Hilbert's Eighth and Tenth Problems 116
6.5 Yet another universal equation 122
6.6 Yet another Diophantine set with non-Diophantine complement 123

7 Decision Problems in Number Theory 129
7.1 The number of solutions of Diophantine equations 129

7.2 Non-effectivizable estimates in the theory of exponential Diophantine equations 130

7.3 Gaussian integer counterpart of Hilbert's Tenth Problem 138

7.4 Homogeneous equations and rational solutions 146

8 Diophantine Complexity 153

8.1 Principal definitions 153

8.2 A bound for the number of unknowns in exponential Diophantine representations 156

9 Decision Problems in Calculus 165

9.1 Diophantine real numbers 165

9.2 Equations, inequalities, and identities in real variables 168

9.3 Systems of ordinary differential equations 174

9.4 Integrability 177

10 Other Applications of Diophantine Representations 181

10.1 Diophantine games 181

10.2 Generalized knights on a multidimensional chessboard 184

Appendix 199

1 The Four Squares Theorem 199

2 Chinese Remainder Theorem 200

3 Kummer's Theorem 201

4 Summation of a generalized geometric progression 202

Hints to the Exercises 205

Bibliography 221

List of Notation 257

Name Index 259

Subject Index 263

Series Foreword

Theoretical computer science has now undergone several decades of development. The "classical" topics of automata theory, formal languages, and computational complexity have become firmly established, and their importance to other theoretical work and to practice is widely recognized. Stimulated by technological advances, theoreticians have been rapidly expanding the areas under study, and the time delay between theoretical progress and its practical impact has been decreasing dramatically. Much publicity has been given recently to breakthroughs in cryptography and linear programming, and steady progress is being made on programming language semantics, computational geometry, and efficient data structures. Newer, more speculative, areas of study include relational databases, VLSI theory, and parallel and distributed computation. As this list of topics continues expanding, it is becoming increasingly important that the most significant work be distilled and communicated in a manner that will facilitate further research and application of this work. By publishing comprehensive books and specialized monographs on the theoretical aspects of computer science, the series on Foundations of Computing provides a forum in which important research topics can be presented in their entirety and placed in perspective for researchers, students, and practitioners alike.

Michael R. Garey
Albert R. Meyer

A Note on the Translation

The English version of this book is a translation from the Russian, but without an official translator. A complete English first draft was prepared by the author, Yuri Matiyasevich, himself. The task of transforming this draft into a reasonably polished English text was undertaken by David Jones of MIT Press and myself, neither of us knowing any Russian. David began by translating Yuri's computer files into TEX, specifically AMS-LATEX, and by helping me to set up this dialect of TEX in my computer. As David completed each chapter, he passed it on to me. After I had made a first pass it would go to David and to Yuri for multiple iterations of the process. In the process, Yuri also added numerous items to the bibliography that were not present in the Russian edition.

David and I were in almost daily contact by electronic mail for close to a year, joined by Yuri when the communication channels permitted. Although David and I have never met, I think we know one another rather well by now. Many were the arguments we had over the need for a comma, and never again will I write the word "which" without wondering whether David would permit me to use it rather than "that" in the given context.

The Russian name of the discipline referred to in English as *recursion theory* or *computability theory* translates literally as the *theory of algorithms.* In current useage however, that English phrase suggests a rather different subject, namely algorithmic analysis. For this reason, we decided to use the English terminology. In all other respects, we have tried to stay as close to the original as possible.

Throughout the book, we have transliterated Cyrillic text according to the current conventions of the AMS *Mathematical Reviews.* The only exceptions are names of people who have a clear preference for an alternative spelling.

Martin Davis

Foreword

While I was still an undergraduate at City College in New York, I read my teacher E. L. Post's plaint that Hilbert's Tenth Problem "begs for an unsolvability proof." This was the beginning of my lifelong obsession with the problem. Although I have had the good fortune to be able to make some contributions towards the "unsolvability proof" for which the problem was begging, my greatest insight turned out to be a thought I had uttered in jest. During the 1960s I often had occasion to lecture on Hilbert's Tenth Problem. At that time it was known that the unsolvability would follow from the existence of a single Diophantine equation that satisfied a condition that had been formulated by Julia Robinson. However, it seemed extraordinarily difficult to produce such an equation, and indeed, the prevailing opinion was that one was unlikely to exist. In my lectures, I would emphasize the important consequences that would follow from either a proof or a disproof of the existence of such an equation. Inevitably during the question period I would be asked for my own opinion as to how matters would turn out, and I had my reply ready: "I think that Julia Robinson's hypothesis is true, and it will be proved by a clever young Russian."

This book was written by that Russian. In 1970, Yuri Matiyasevich presented his beautiful and elegant construction of a Diophantine equation that satisfies Julia Robinson's hypothesis. This showed not only that Hilbert's Tenth Problem is unsolvable, but also that two fundamental concepts arising in different areas of mathematics are equivalent. The notion of *recursively enumerable* or *semidecidable* set of natural numbers from computability theory turns out to be equivalent to the purely number-theoretic notion of *Diophantine* set. Dr. Matiyasevich has taken full advantage of the rich interplay between the methods of elementary number theory and computability theory that this equivalence makes possible to produce a remarkable and appealing book. The reader will find new and simplified proofs of some of the main results, various extensions and applications, and many interesting exercises.

The history of the subject is recounted with meticulous care in the "Commentaries" that follow each chapter of the book. Dr. Matiyasevich has also provided a very personal account of his involvement with Hilbert's Tenth Problem in his article "My Collaboration with Julia Robinson" in the *Mathematical Intelligencer* (Matiyasevich [1992]). In this brief introduction, I would like to offer a few vignettes from my own involvement with the problem. As a graduate student at Princeton University, I had chosen what I knew was an excellent topic for my dissertation: the extension of Kleene's arithmetic hierarchy into the constructive transfinite, what has come to be called the *hyperarithmetic hierarchy*. This was a completely unexplored area, was quite fascinating, and was sure to yield results. But, I couldn't stop myself from thinking about Hilbert's Tenth Problem. I thought it unlikely that I would get anywhere on such a difficult problem and tried without success to discipline myself to stay away from it. In the end, my dissertation, written under the supervision of Alonzo Church, had results on both the hyperarithmetic hierarchy and Hilbert's Tenth Problem. In my dissertation, I conjectured the equivalence of the two notions mentioned above (in this book referred to as my "daring hypothesis") and saw how to improve Gödel's use of the Chinese Remainder Theorem as a coding device so as to obtain a representation for recursively enumerable sets that formally speaking seemed close to the desired result. The obstacle that remained in this so-called Davis normal form was a single bounded universal quantifier.

I met Julia Robinson at the 1950 International Congress of Mathematicians in Cambridge, Massachusetts, immediately after completing my doctorate. She had approached Hilbert's Tenth Problem from a direction opposite to mine. Where I had tried to simplify the arithmetic representation of arbitrary recursively enumerable sets, she had been trying to produce Diophantine definitions for various specific sets and especially for the exponential function. She had introduced what was to become her famous "hypothesis" and shown that under that assumption the exponential function is in fact Diophantine. It's been said that I told her that I doubted that her approach would get very far, surely one of the more foolish statements I've made in my life.

During the summer of 1957, there was an intensive five week "Institute for Logic" at Cornell University attended by almost all American logicians. Hilary Putnam and I together with our families were sharing a house in Ithaca, and he and I began collaborating, almost without thinking about it. Hilary proposed the idea of using the Chinese Remainder Theorem coding one more time to code the sequences whose existence was asserted by the bounded universal quantifier in the Davis normal form. My first reaction was skeptical. But, as pointed out in the Commentary to Chapter 3 in this book, the Chinese Remainder Theorem provides a "unique opportunity" because of the fact that polynomials preserve congruences. In fact, we were able to obtain two particular sets with quite simple definitions concerning which we were able to show that their being Diophantine would imply the same for all recursively enumerable sets.

Hilary and I resolved to seek other opportunities to work together, and we were able to obtain support for our research during the three summers of 1958, 1959, and 1960. We had a wonderful time. We talked constantly about everything under the sun. Hilary gave me a quick course in classical European philosophy, and I gave him one in functional analysis. We talked about Freudian psychology, about the current political situation, about the foundations of quantum mechanics, but mainly we talked mathematics. It was during the summer of 1959 that we did our main work together on Hilbert's Tenth Problem. In a recent letter, Hilary wrote:

> What I remember from that summer is not so much the mathematical details as the sheer *intensity* with which we worked. I have never in my life been so absorbed in a mathematical problem, and I'm sure the same was true of you. Our method, as I remember it, was that one of us would propose an attack and we would both work on it together, writing on the board and arguing with each other, making suggestions, etc., until something came of it or we reached a dead end. I could not let go of the problem even at night; this is the only time when I regularly stayed up to four in the morning ... I think we felt in our bones that the problem would yield to our approach; otherwise I can't explain the sense of mounting excitement.

Our "approach" was still to apply the Chinese Remainder Theorem to Davis normal form. But this time, we were combining this attack with Julia Robinson's methods, attempting to see if by permitting exponentiation in our Diophantine definitions we could eliminate the troublesome bounded universal quantifier. The problem in using the Chinese Remainder Theorem was the need for suitable moduli, relatively prime in pairs. Gödel's method was to obtain such moduli in an arithmetic progression, and hence definable in Diophantine terms. We found ourselves with the need to find exponential Diophantine definitions for sums of the reciprocals of the terms of a finite arithmetic progression as well as of the product of such terms. To deal with the second problem, we used binomial coefficients with rational numerators, for which we could find exponential Diophantine definitions extending Julia Robinson's methods, but requiring the binomial theorem with rational exponents, an infinite power series expansion. For the first, we used a rather elaborate (and as it turned out, quite unnecessary) bit of elementary analysis, involving the Taylor expansion of the Gamma function. Even with all that, we still couldn't get the full result we wanted. We needed to be able to assert that if one of our moduli was a divisor of a product that it had to necessarily divide one of the factors. And this seemed to require that the moduli be not only relatively prime in pairs, but actual prime numbers. In the end, we were forced to assume the hypothesis (still unproved to this date) that there are arbitrarily long arithmetic progressions of prime numbers, in order to prove that every recursively enumerable set has an exponential Diophantine definition.

We sent our results to Julia Robinson, and she responded shortly thereafter saying:

> I am very pleased, surprised, and impressed with your results on Hilbert's Tenth Problem. Quite frankly, I did not think your methods could be pushed further ...
>
> I believe I have succeeded in eliminating the need for [the assumption about primes in arithmetic progression] by extending and modifying your proof. I have this written out for my own satisfaction but it is not yet in shape for anyone else.

The letter also showed quite neatly how to dispense with the messy analysis involving the Gamma function that Hilary and I had used. Soon afterwards, we received the details of Julia's proof, and it was our turn to be "very pleased, surprised, and impressed." She had avoided our hypothesis about primes in arithmetic progression in an elaborate and very clever argument by making using of the prime number theorem for arithmetic progressions to obtain enough primes to permit the proof to go through. She graciously accepted our proposal that our work (which had already been submitted for publication) be withdrawn in favor of a joint publication. Soon afterwards, she succeeded in a drastic simplification of the proof: where Hilary and I were trying to use the Gödel coding to obtain a logical equivalence, her elegant argument made use of the fact that the primes were only needed for the implication in one direction, and that in that direction one could make do with a prime divisor of each modulus. (Later Yuri Matiyasevich showed that in fact any sufficiently large coprime moduli could be used so that our efforts in connection with prime factors were really unnecessary; see Exercise 2 in Chapter 6.)

With the result that every recursively enumerable set has an exponential Diophantine definition combined with Julia Robinson's earlier work on Diophantine definitions of the exponential function, it was now clear that my "daring hypothesis" of the equivalence of the two notions, recursively enumerable set and Diophantine set, was entirely equivalent to the much weaker hypothesis (now called JR) that Julia Robinson had proposed ten years earlier that one single Diophantine equation could be found whose solutions satisfied a simple condition. During the summer of 1960, Hilary and I were in Boulder, Colorado participating in a special institute intended to teach mathematicians something about physics. Hilary and I continued to argue about quantum mechanics and explored the possibility of finding a third degree equation to satisfy Julia Robinson's condition. It turned out once again that we needed information that the number theorists were unable to provide, this time about the units in pure cubic extensions of the rational numbers.

During the following years, I continued trying to prove Julia Robinson's hypothesis. I was particularly interested in trying to use what was known about quadratic number fields. It was this work that led me to the equation $9(x^2 + 7y^2)^2 - 7(u^2 + 7v^2)^2$, in which there is still some interest. (See the Commentary to Chapter 2.) At this time, Julia had become rather pessimistic about JR and, for a brief period, she actually worked towards a positive solution of Hilbert's Tenth Problem. A letter from her dated April 1968 responding to my report on the above equation said:

> I have enjoyed studying it, but my faith in JR still hasn't been restored. However, for the first time. I can see how it might be proved. Indeed, maybe your equation works, but it seems to need an infinite amount of good luck!

Early in 1970, a telephone call from my old friend Jack Schwartz informed me that the "clever young Russian" I had predicted had actually appeared. Julia Robinson sent me a copy of John McCarthy's notes on a talk that Grigoriĭ Tseitin had given in Novosibirsk on the proof by the twenty-two-year-old Yuri Matiyasevich of the Julia Robinson hypothesis. Although the notes were brief, everything important was there, and I was able to have the great pleasure of reconstructing the proof. But I was not satisfied until I had produced my own variant of Dr. Matiyasevich's proof and presented it (on March 10) at a seminar at Rockefeller University at Hao Wang's invitation.

I met Yuri a few months later at the International Congress of Mathematicians in Nice, where he was an invited speaker. I was finally able to tell him that I had been predicting his appearance for some time.

Martin Davis

Preface to the English Edition

I am happy to see the English translation of this book being published in the same year as the Russian original. This became possible thanks to the initiative of Albert Meyer, who proposed the publication of the translation at a time when I had not finished rewriting the Russian original for the $(n+1)$st time. I am very grateful to him for this.

The Russian original was prepared with the aid of ChiWriter, a system that is friendly both to mathematics and to Cyrillic text. However, I continued to write the book in the old style, i.e., with pen and paper, and my wife Nina had the patience to type at least half of the book.

When the time came to prepare the first draft translation in TEX, it turned out that the ChiWriter-to-TEX converter did not understand many complicated formulas, and it was the skill and experience of David Jones that came to help. He and Martin Davis, one of the pioneers of Hilbert's Tenth Problem, did a great job trying to convey into English all the nuances of Russian phrases. They also revealed many misprints and errors but of course the responsibilty for the (surely) still-remaining ones is entirely mine. I express my gratitude to both of them.

Yuri Matiyasevich

Preface

In 1900 David Hilbert [1900] delivered his famous lecture entitled "Mathematische Probleme" before the Second International Congress of Mathematicians. This paper contains 23 problems, or, more precisely, 23 groups of related problems, that the nineteenth century left for the twentieth century to solve. Problem number ten is about Diophantine equations:

> 10. DETERMINATION OF THE SOLVABILITY OF A DIOPHANTINE EQUATION
> Given a diophantine equation with any number of unknown quantities and with rational integral numerical coefficients: *To devise a process according to which it can be determined by a finite number of operations whether the equation is solvable in rational integers.*

Today we read the words "devise a process" to mean "find an algorithm." When Hilbert's Problems were posed, there was no mathematically rigorous *general* notion of algorithm available. The lack of such a notion was not in itself an obstacle to a positive solution of Hilbert's Tenth Problem, because for any *particular* algorithm it was always clear that it actually gave the desired general method for solving the corresponding problem.

During the 1930's, Kurt Gödel, Alonzo Church, Alan Turing, and other logicians provided a rigorous formulation of the notion of *computability*; this made it possible to establish *algorithmic unsolvability*, i.e., the impossibility of the existence of an algorithm with certain properties. Soon afterwards the first examples of algorithmically unsolvable problems were found, first in mathematical logic itself and then in other branches of mathematics.

Computability theory produced all the necessary tools for tackling the unsolvability of Hilbert's Tenth Problem. The first in a series of papers in this direction appeared at the beginning of the 1950's. The continuing effort culminated in a "negative solution" of Hilbert's Tenth Problem in 1970.

As in the case of many other problems whose solution was long awaited, the technique developed for the resolution of Hilbert's Tenth Problem is of independent value because it has other applications. Some of these are rather striking, and taken together, these other applications are perhaps even more important than the solution of Hilbert's Tenth Problem. The main technical result implying the unsolvability of Hilbert's Tenth Problem asserts that the class of *Diophantine sets* is identical with the class of *recursively enumerable sets.* Another corollary of this same result that uses no special terminology states: *It is possible to exhibit explicitly a polynomial with integer coefficients such that the set of all the positive values it assumes for integer values of its variables is exactly the set of all prime numbers.*

The present book is devoted to the algorithmic unsolvability of Hilbert's Tenth Problem and related topics; the numerous partial results that have been obtained in the direction of a positive solution of the Problem are hardly considered.

The negative solution of Hilbert's Tenth Problem is presented (in varying detail) in many books and surveys, in particular, in the following publications: Azra [1971], Börger [1989], Davis [1973a], Fenstad [1971], Hermes [1972, 1978], Hirose [1973], Jones and Matiyasevich [1991], Kaplansky [1977], Manin [1973], Manin and Panchishkin [1990], Margenstern [1981], Matiyasevich [1972a], Mijajlović, Marković, and Došen [1986], Ruohonen [1972, 1980], Salomaa [1985], Smoryński [1991], Sussman [1971], and Zakharov [1970].

A distinguishing feature of this book is that, in addition to presenting the negative solution of the problem, it contains a number of diverse applications of the technique developed for that solution. At present these applications are scattered among various publications, mainly journal articles. During the twenty years that have passed since the problem was "unsolved," many improvements and modifications of the original proof have been suggested. In addition to these, the book contains several new, previously unpublished proofs.

Understanding the negative solution of Hilbert's Tenth Problem naturally requires some knowledge from both number theory and mathematical logic. Wishing to make the book accessible to a broader readership, especially to younger mathematicians, the author has tried to reduce the mathematical prerequisites needed by the reader. In particular, no knowledge of computability theory is presupposed. All the necessary notions are defined in the book, and thus it can serve as an introduction to this fascinating subject (but of course the book cannot be used for its systematic study). A few number-theoretical results that are usually not part of the basic mathematical curriculum are proved in the *Appendix*.

Each chapter is concluded by *Exercises*, which contain problems of varying difficulty. While some of them are quite elementary, others amount to small research problems. The aim of the exercises is to present diverse results, but without proofs. Of course, this division (on the one hand, the main content of the book presented with full proofs and, on the other, the results relegated to the exercises) represents a rather subjective judgement by the author. In particular, the exercises contain results requiring special knowledge or having cumbersome proofs as well as results that are, most likely, far from the best possible or that were thought to be of limited interest. The *Hints* supply the ideas of the proofs and/or references to the literature.

In addition to the exercises, there are a few *Open Questions* and *Unsolved Problems.* Again, the division is quite subjective. It may be that an open question has not been settled simply because no one has tackled it seriously, and the answer may turn out to be without significance. On the other hand, the unsolved problems have attracted the attention of many capable researchers, and solving these problems may require decades.

Each chapter is completed by a *Commentary* providing a historical view of its contents. This seems to be worthwhile because the logical order of presentation used in the book often doesn't coincide with the chronological order in which the results had originally been obtained.

The *Bibliography* lists all of the principal publications concerned with the negative solution to Hilbert's Tenth Problem as well as most of the papers that apply the technique developed for obtaining that solution. The author would be grateful if omissions were called to his attention.

The book need not be read consecutively. It may be said to consist of two parts. The first part, consisting of Chapters 1–5, presents the solution of Hilbert's Tenth Problem. Figure 1 exhibits the dependencies among the sections of the first part. To determine which sections should be read before a particular section U.V, locate that section in the figure and imagine vertical and horizontal coordinate axes placed on the page in such a manner that their point of intersection is at the period in U.V. Then, before reading section U.V you will need to read all sections lying in the upper left quadrant including its boundary.

Similarly, Figure 2 shows how the sections of the second part, devoted to applications, depend on the sections from the first part (but not dependencies between sections of the first part and not dependencies between sections of the second part):

There are few dependencies among the sections of the second part. Sections 6.1–6.3 offer three different techniques for achieving the same aim; it suffices to know any one of them in order to read Sections 6.4–6.6 and 9.1. Similarly, Sections 4.5 and 6.5 present two different constructions of a universal equation, either of which can serve as background for reading Sections 6.6 and 8.1. Section 8.2 presupposes acquaintance with Section 7.2, which in turn presupposes knowledge of Sections 6.2–6.3; Section 9.4 uses results from Section 9.2, and Section 10.1 is based on Section 6.6.

					5.1	
					5.2	
1.1						
1.2						
1.3						
1.4						
1.5				3.1	5.3	5.4
	1.6	2.1				
		2.2				
		2.3				
		2.4				
		2.5				
3.1	3.2	3.3	3.4			
			3.5	3.6	5.5	
4.1						
4.2		4.3	4.4			
			4.5			
			4.6			5.6
						5.7

Figure 1

2.1	2.4	3.4	5.4	5.5	5.7	**7.1**	**7.2**
7.3	**6.4**	**6.2**		**6.1**	**9.1**	**9.2**	**9.3**
7.4		**6.3**			**10.1**	**10.2**	

Figure 2

Hilbert's Tenth Problem

1 Principal Definitions

In this chapter we introduce *Diophantine sets*, the main subject of study in this book, and we establish some of their simplest properties.

1.1 Diophantine equations as a decision problem

Let us recall that a *Diophantine equation* is an equation of the form

$$D(x_1, \dots, x_m) = 0, \tag{1.1.1}$$

where D is a polynomial with integer coefficients. In addition to (1.1.1), Diophantine equations can be written in the more general form

$$D_{\mathrm{L}}(x_1, \dots, x_m) = D_{\mathrm{R}}(x_1, \dots, x_m), \tag{1.1.2}$$

where D_{L} and D_{R} are again polynomials with integer coefficients. When we speak of "an arbitrary Diophantine equation," we shall have in mind an equation of the form (1.1.1) since an equation of the form (1.1.2) can easily be transformed into an equation of the form (1.1.1) by transposing all the terms to the left-hand side. However, we will often use the notation (1.1.2) for particular equations if this form turns out to be easier to grasp. We shall also take another advantage of the more general form (1.1.2); namely, in this case we can demand that D_{L} and D_{R} are polynomials with non-negative coefficients.

Diophantine equations typically have several unknowns, and we must distinguish the *degree of* (1.1.1) *with respect to a given unknown* x_i and the (*total*) *degree of* (1.1.1), i.e., the maximum, over all the monomials constituting the polynomial D, of the sum of the degrees of the individual variables in such a monomial.

In specifying a Diophantine equation it is necessary to provide not only a representation of the form (1.1.1) (or equivalently (1.1.2)), but also to indicate the range of the unknowns. Hilbert, in his Tenth Problem, spoke of solutions *in rational integers*. In the present book we shall just use the term "*integers*" because the case of *algebraic integers* will be almost completely ignored. (The cases of other ranges for the unknowns will be considered in Sections 1.3, 7.3, and 7.4.)

Hilbert's Tenth Problem is an example of a *decision problem*. A decision problem consists of countably many *individual problems*, for each of which an answer "YES" or "NO" is to be given. We shall refer to these individual problems as *subproblems* of the corresponding decision problem. Each individual subproblem is specified by a finite amount of information. (In the case of Hilbert's Tenth Problem, this information is the polynomial D from (1.1.1)).

The essence of a decision problem lies in the requirement to find a single method that can be used for obtaining the answer to any of the individual subproblems. Since Diophantus's time, number theorists have found solutions for many Diophantine equations and have established the unsolvability of many other equations. However, for many particular classes of equations, and even for certain individual equations, it was necessary to invent a specific method. In the Tenth Problem, Hilbert asked for a *universal* method for deciding the solvability of Diophantine equations.

A decision problem can be solved either directly or indirectly. In the the former case, one provides a procedure for finding the answer to any individual subproblem, while in the latter case one *reduces* the given decision problem to another, the solvability of which has already been shown. We will not give a formal definition of what constitutes such a reduction, because the general theory of reducibility will not be needed, and in each particular case of such a reduction it will be clear from the context what is actually meant.

An unsolvability proof for a decision problem can also be either direct or indirect. In the latter case one also reduces one problem to another, but what is required is a reduction in the reverse direction. Namely, to establish the unsolvability of a decision problem, one has to reduce *to it* another problem the unsolvability of which has already been shown. In the first few chapters of this book, we reduce to Hilbert's Tenth Problem successively more and more complicated problems. This chain of reductions should lead ultimately to a problem for which we can give a direct proof of unsolvability. However, to be able to give such a proof, we need some way of surveying all conceivable methods of solving the problem. The possibility of doing this became available only after the development of the mathematically rigorous general notion of computability. All the necessary definitions are given in Chapter 5, where we establish the algorithmic unsolvability of Hilbert's Tenth Problem.

1.2 Systems of Diophantine equations

Although Hilbert asked for a general method for deciding whether a single Diophantine equation does or does not have a solution, Diophantus had considered systems of equations as well. However, it easy to see that a positive solution of Hilbert's Tenth Problem would also give us a method for deciding whether a system of Diophantine equations does or does not have a solution. In fact, the system consisting of the k equations

$$\begin{aligned} D_1(x_1, \ldots, x_m) &= 0 \\ &\vdots \\ D_k(x_1, \ldots, x_m) &= 0 \end{aligned} \tag{1.2.1}$$

has a solution in integers $x_1, \ldots, x_m$ if and only if the Diophantine equation

$$D_1^2(x_1, \ldots, x_m) + \cdots + D_k^2(x_1, \ldots, x_m) = 0 \tag{1.2.2}$$

has one. Moreover, the set of solutions of (1.2.1) coincides with the set of solutions of (1.2.2). Thus, for systems of Diophantine equations the number of equations is not such an essential characteristic as it is in the cases of linear algebraic or differential equations.

In what follows, we shall also use a transformation in the reverse direction, namely, the transformation of an equation

$$D(x_1, \ldots, x_m) = 0 \tag{1.2.3}$$

into the system of Diophantine equations

$$\begin{aligned} D_1(x_1, \ldots, x_m, y_1, \ldots, y_n) &= 0 \\ &\vdots \\ D_k(x_1, \ldots, x_m, y_1, \ldots, y_n) &= 0 \end{aligned} \tag{1.2.4}$$

that, in general, contains some additional unknowns $y_1, \ldots, y_n$. The operation of passing from (1.2.3) to (1.2.4) need not be inverse to that of passing from (1.2.1) to (1.2.2); i.e., if we compress the system (1.2.4) into a single equation in the manner described above we would not in general obtain the initial equation (1.2.3). The only relation between (1.2.3) and (1.2.4) that is of interest to us is that system (1.2.4) should have a solution if and only if equation (1.2.3) has one. We require neither that every solution of (1.2.3) could be expanded (by choosing values of $y_1, \ldots, y_n$) into a solution of (1.2.4), nor that every solution of (1.2.4) should include a solution of (1.2.3).

One possible reason for transforming an equation (1.2.3) into a system (1.2.4) could be to obtain a system consisting of very simple equations. For example, it is easy to see that any Diophantine equation can be transformed into an equivalent system (in the sense explained above) consisting of equations of the two forms

$$\alpha = \beta + \gamma \tag{1.2.5}$$

and

$$\alpha = \beta\gamma, \tag{1.2.6}$$

where α, β, and γ are either particular natural numbers or certain of the unknowns $x_1, \ldots, x_m, y_1, \ldots, y_n$. We proceed to explain how such a transformation is possible using the equation

$$4x^3y - 2x^2z^3 - 3y^2x + 5z = 0 \tag{1.2.7}$$

as an example. First of all we transpose the negative terms, obtaining the equation

$$4x^3y + 5z = 2x^2z^3 + 3y^2x. \tag{1.2.8}$$

Next we introduce 14 new unknowns and obtain the equivalent system

$$\begin{gathered} p_1 = 4x, \quad p_2 = p_1x, \quad p_3 = p_2x, \quad p_4 = p_3y; \\ q_1 = 5z; \\ r_1 = 2x, \quad r_2 = r_1x, \quad r_3 = r_2z, \quad r_4 = r_3z, \quad r_5 = r_4z; \\ s_1 = 3y, \quad s_2 = s_1y, \quad s_3 = s_2x; \\ t_1 = p_4 + q_1, \quad u_1 = r_5 + s_3, \quad t_1 = 1u_1. \end{gathered} \tag{1.2.9}$$

As an application of this simple transformation technique, let us consider what would result if we first transform a Diophantine equation into an equivalent system (1.2.1) consisting of equations of the forms (1.2.5) and (1.2.6), and then compress this system into a single equation (1.2.2). Clearly, the initial equation has or hasn't a solution according as the resulting equation (1.2.2) has or hasn't one. However, the latter equation is of degree 4 regardless of the degree of the former. Thus, *to solve Hilbert's Tenth Problem positively, it would be sufficient to find a method for deciding whether a Diophantine equation of degree* 4 *has a solution.*

1.3 Solutions in natural numbers

In his Tenth Problem, Hilbert asked about solutions in integers. Sometimes an integer solution is evident. For example, the equation

$$(x+1)^3 + (y+1)^3 = (z+1)^3 \tag{1.3.1}$$

clearly has infinitely many solutions of the form $x = z, \quad y = -1$. On the other hand, the fact that equation (1.3.1) has no solutions in non-negative x, y, and z

is not trivial at all. Thus, for a particular Diophantine equation, *the problem of deciding whether it has an integer solution* and *the problem of deciding whether it has a non-negative solution* are in general two separate problems.

On the other hand, let

$$D(x_1, \dots, x_m) = 0 \tag{1.3.2}$$

be an arbitrary Diophantine equation and suppose that we are looking for its non-negative solutions. Let us consider the following system:

$$\begin{aligned} D(x_1, \dots, x_m) &= 0 \\ x_1 &= y_{1,1}^2 + y_{1,2}^2 + y_{1,3}^2 + y_{1,4}^2 \\ &\vdots \\ x_m &= y_{m,1}^2 + y_{m,2}^2 + y_{m,3}^2 + y_{m,4}^2. \end{aligned} \tag{1.3.3}$$

Clearly, any solution of this system in *arbitrary* integers includes a solution of (1.3.2) in *non-negative* integers. The converse is true as well; namely, for any solution of (1.3.2) in non-negative integers $x_1, \dots, x_m$, there are integer values of $y_{1,1}, \dots, y_{m,4}$ giving a solution of (1.3.3), because every non-negative integer can be represented as the sum of the squares of four integers (see the Appendix). As we know from Section 1.2, the system (1.3.3) can be compressed into a single equation

$$E(x_1, \dots, x_m, y_{1,1}, \dots, y_{m,4}) = 0 \tag{1.3.4}$$

solvable in integers if and only if the original equation (1.3.2) is solvable in non-negative integers.

Thus, we have shown that the decision problem of determining the existence or non-existence of non-negative solutions is reducible to the decision problem of determining the existence or non-existence of integer solutions. Hence we have proved that to establish the unsolvability of Hilbert's Tenth Problem in its original form, it is sufficient to establish the unsolvability of its analog for non-negative solutions. For technical reasons, it is easier to work with non-negative solutions, and in the rest of the book lower-case italic Latin letters will denote non-negative integers unless the contrary is clearly stated. Following the tradition in mathematical logic, we will also speak of the non-negative integers as the *natural numbers*; thus in particular, we shall consider 0 to be a natural number.

Our further efforts will be directed towards establishing the unsolvability of the analog of Hilbert's Tenth Problem for natural number solutions. We shall reach this goal in Chapter 5. However, *a priori*, it is conceivable that this analog is solvable whereas the original problem is unsolvable. Let us check that in fact this cannot happen, i.e., that we lose essentially nothing by restricting the range of unknowns to natural numbers.

Let

$$D(\chi_1, \ldots, \chi_m) = 0 \tag{1.3.5}$$

be an arbitrary Diophantine equation and suppose that now we are seeking its solutions in integers $\chi_1, \ldots, \chi_m$. Let us consider the equation

$$D(x_1 - y_1, \ldots, x_m - y_m) = 0. \tag{1.3.6}$$

Clearly, any solution of equation (1.3.6) (which, according to our agreement concerning lower-case Latin letters, consists of natural numbers $x_1, \ldots, x_m, y_1, \ldots, y_m$) generates a solution

$$\begin{aligned} \chi_1 &= x_1 - y_1 \\ &\vdots \\ \chi_m &= x_m - y_m \end{aligned} \tag{1.3.7}$$

of equation (1.3.5) in integers $\chi_1, \ldots, \chi_m$. On the other hand, for any solution $\chi_1, \ldots, \chi_m$ of equation (1.3.5) one can find natural numbers $x_1, \ldots, x_m, y_1, \ldots, y_m$ satisfying (1.3.7) and hence giving a solution of equation (1.3.6).

Thus, we have also made a reduction in the opposite direction; i.e., we have shown that the problem of the existence of integer solutions is reducible to the problem of the existence of solutions in natural numbers. The two reductions show that these two problems are equivalent *as decision problems* although, as was discussed at the beginning of this section, for a particular equation the answer may well depend on the range of the unknowns.

1.4 Families of Diophantine equations

In addition to systems of Diophantine equations, number theory also studies families of Diophantine equations. By a *family of Diophantine equations*, we understand a relation of the form

$$D(a_1, \ldots, a_n, x_1, \ldots, x_m) = 0 \tag{1.4.1}$$

where D is a polynomial with integer coefficients with respect to all the variables $a_1, \ldots, a_n, x_1, \ldots, x_m$, separated into *parameters* $a_1, \ldots, a_n$ and *unknowns* $x_1, \ldots, x_m$. Fixing values of the parameters results in the particular Diophantine equations that comprise the family. A family of Diophantine equations is *not* an infinite *system* of equations because the unknowns need not satisfy all the equations simultaneously, as would be the case for a system. Families of Diophantine equations are also called *parametric equations.* For a parametric Diophantine equation, we must distinguish the *degree of the equation with respect to its unknowns* from the *degree of the equation with respect to all of its variables.*

For different values of the parameters, one can obtain equations that do have solutions as well as equations that do not. The parametric Diophantine equation (1.4.1) defines a set $\mathfrak{M}$ consisting of the n-tuples of values of the parameters $a_1, \ldots, a_n$ for which there are values of the unknowns $x_1, \ldots, x_m$ satisfying (1.4.1):

$$\langle a_1, \ldots, a_n \rangle \in \mathfrak{M} \iff \exists x_1 \ldots x_m \, [D(a_1, \ldots, a_n, x_1, \ldots, x_m) = 0]. \quad (1.4.2)$$

The number n is called the *dimension* of the set $\mathfrak{M}$, and equivalence (1.4.2) is called a *Diophantine representation* of $\mathfrak{M}$; taking some liberty, one can speak of (1.4.1) as well as (1.4.2) as a Diophantine representation. Sets having Diophantine representations are also called *Diophantine.* Clearly, every Diophantine set has infinitely many Diophantine representations.

Diophantine sets are the main object of study of this book. In number theory an equation is usually the primary object and one looks for a description of the corresponding set in some other terms. We shall mainly deal with problems of the opposite kind: *A set of n-tuples of natural numbers is given. Is this set Diophantine? If it is, find a Diophantine representation for it.*

Sometimes it evident that a set is Diophantine, for example, the set of all even numbers. For other sets it need not be so evident, and in some cases it can be technically difficult to prove that a set is in fact Diophantine. As an example, we can mention the set of all prime numbers, which will be shown to be Diophantine in Section 3.4.

In addition to determining whether some particular sets are Diophantine, one might seek an alternative characterization of the class of Diophantine sets. Such a characterization in terms of the notions of computability theory will be given in Chapter 5, and this will permit us to obtain the negative solution of Hilbert's Tenth Problem.

In this chapter we shall establish some of the simplest properties of Diophantine sets and exhibit several examples of Diophantine sets important for the rest of the book.

It is easy to see that *the union of two Diophantine sets of the same dimension is also Diophantine*. In fact, if

$$D_1(a_1, \ldots, a_n, x_1, \ldots, x_{m_1}) = 0 \tag{1.4.3}$$

and

$$D_2(a_1, \ldots, a_n, x_1, \ldots, x_{m_2}) = 0 \tag{1.4.4}$$

are Diophantine representations of two sets, then the equation

$$D_1(a_1, \ldots, a_n, x_1, \ldots, x_{m_1}) \cdot D_2(a_1, \ldots, a_n, x_1, \ldots, x_{m_2}) = 0 \tag{1.4.5}$$

is a Diophantine representation of their union.

The intersection of two Diophantine sets of the same dimension is Diophantine as well; namely, it is defined by the equation

$$D_1^2(a_1, \ldots, a_n, x_1, \ldots, x_{m_1}) + D_2^2(a_1, \ldots, a_n, y_1, \ldots, y_{m_2}) = 0. \tag{1.4.6}$$

Note that we not only used the technique of Section 1.2 for combining the two equations (1.4.3) and (1.4.4) into a single equation, but we also first changed the unknowns in equation (1.4.4).

The complement of a Diophantine set of n-tuples of natural numbers (with respect to the set of all n-tuples of natural numbers) need not be Diophantine, but this fact is not trivial and will be proved only in Section 4.6.

In the definition of a Diophantine representation (1.4.2), no restrictions were imposed on the type of the corresponding equation (1.4.1). We know from Section 1.2 that we could require (1.4.1) to be an equation of degree 4. In what follows, we shall take advantage of another possibility for specializing the equation in a Diophantine representation of a set of natural numbers (i.e., in the case $n = 1$). Namely, it is easy to see that the equation

$$D(a, x_1, \ldots, x_m) = 0 \tag{1.4.7}$$

has a solution in unknowns $x_1, \ldots, x_m$ if and only if the equation

$$(x_0 + 1)(1 - D^2(x_0, \ldots, x_m)) - 1 = a \tag{1.4.8}$$

has a solution in unknowns $x_0, \ldots, x_m$. In fact, any solution of equation (1.4.7) can be expanded to a solution of (1.4.8) by putting $x_0 = a$; on the other hand, in any solution of equation (1.4.8) the factor $1 - D^2(x_0, \ldots, x_m)$ must be positive, which is possible only if $D(x_0, \ldots, x_m) = 0$. But then (1.4.8) implies that $x_0 = a$ and hence that the equality (1.4.7) also holds.

Thus, *a set of natural numbers is Diophantine if and only if it is the set of all natural number values assumed by some polynomial with integer coefficients for natural number values of its variables.*

1.5 Logical terminology

The entire contents of this book could be presented in terms of Diophantine sets, their unions and intersections, and the relation of set membership. However, it is often more convenient to use, instead of the language of sets, an essentially equivalent language of properties and relations. For example, instead of saying that the set of even numbers is Diophantine, one can say that the property *is an even number* is Diophantine. Similarly, instead of considering the set with the representation

$$(a_1 - a_2)^2 = x + 1, \tag{1.5.1}$$

one can say that the relation $\neq$ is Diophantine.

More formally, we say that a property $\mathcal{P}$ of natural numbers is a *Diophantine property* if the set of numbers having this property is Diophantine. Correspondingly, an equivalence of the form

$$\mathcal{P}(a) \iff \exists x_1 \ldots x_m \, [D(a, x_1, \ldots, x_m) = 0] \tag{1.5.2}$$

is called a *Diophantine representation of property* $\mathcal{P}$. Similarly, a relation $\mathcal{R}$ among n natural numbers is called a *Diophantine relation* if the set of all n-tuples for which the relation holds is Diophantine. Correspondingly, an equivalence of the form

$$\mathcal{R}(a_1, \ldots, a_n) \iff \exists x_1 \ldots x_m \, [D(a_1, \ldots, a_n, x_1, \ldots, x_m) = 0] \tag{1.5.3}$$

is called a *Diophantine representation of relation* $\mathcal{R}$.

Instead of unions of sets, in this terminology one speaks of the logical connective "OR" (disjunction). In other words, if $\mathcal{R}_1$ and $\mathcal{R}_2$ are Diophantine relations (or properties) then the relation (or property) $\mathcal{R}$ such that for all $a_1, \ldots, a_n$

$$\mathcal{R}(a_1, \ldots, a_n) \iff \mathcal{R}_1(a_1, \ldots, a_n) \vee \mathcal{R}_2(a_1, \ldots, a_n) \tag{1.5.4}$$

is also Diophantine. To prove this formally, we should really begin with given Diophantine representations (1.4.3) and (1.4.4) of the relations $\mathcal{R}_1$ and $\mathcal{R}_2$, respectively, and construct from them a Diophantine representation (1.4.5) of the relation $\mathcal{R}$. However, we shall usually omit this step and confine ourselves to representations of the form (1.5.4). We can consider (1.5.4) to be a generalized Diophantine representation of the relation $\mathcal{R}$ provided that we have already shown that the relations $\mathcal{R}_1$ and $\mathcal{R}_2$ are Diophantine.

Similarly, intersection of sets has as its logical counterpart the connective "AND" (conjunction). So, an equivalence of the form

$$\mathcal{R}(a_1,\dots,a_n) \iff \mathcal{R}_1(a_1,\dots,a_n)\ \&\ \mathcal{R}_2(a_1,\dots,a_n) \tag{1.5.5}$$

can be considered to be a generalized Diophantine representation of the relation $\mathcal{R}$ provided, of course, that we have shown that the relations $\mathcal{R}_1$ and $\mathcal{R}_2$ are Diophantine.

Clearly, we can go even further in this direction and regard any formula constructed from parametric Diophantine equations by using, in whatever order, disjunction, conjunction, and existential quantification as constituting a generalized Diophantine representation. Examples of such formulas will be given in the next section. When required, each generalized Diophantine representation can be transformed into the appropriate corresponding canonical form, (1.4.2), (1.5.2), or (1.5.3).

In simplifying proofs that certain sets are Diophantine, the following notion will often be helpful: A function is called a *Diophantine function* if its graph is a Diophantine set. Correspondingly, a *Diophantine representation of a function F* is an equivalence of the form

$$a = F(b_1,\dots,b_n) \iff \exists x_1 \dots x_m\,[D(a,b_1,\dots,b_n,x_1,\dots,x_m) = 0] \tag{1.5.6}$$

where D is a polynomial with integer coefficients.

Now, along with Diophantine equations we can also treat equations of the form

$$P(\mathfrak{t}_1,\dots,\mathfrak{t}_k) = 0 \tag{1.5.7}$$

where P is a polynomial with integer coefficients and $\mathfrak{t}_1, \dots, \mathfrak{t}_k$ are *Diophantine terms*, i.e., expressions constructed in the usual manner from particular natural numbers, variables, the symbols "+", "−", "·", and symbols for Diophantine functions. As when passing from (1.2.7) to (1.2.9), equation (1.5.7) can be transformed (at the expense of introducing new unknowns) into an equivalent system consisting of equations of the forms (1.2.5) and (1.2.6) and of the form

$$\alpha = F(\beta_1,\dots,\beta_n) \tag{1.5.8}$$

where F is a Diophantine function and α and $\beta_1, \ldots, \beta_n$ are variables or particular natural numbers. Furthermore, we can replace each equation of the form (1.5.8) by a copy of equation (1.5.6) with a replaced by α and $b_1, \ldots, b_n$ replaced by $\beta_1, \ldots, \beta_n$ and also $x_1, \ldots, x_m$ replaced by some new variables that have not yet been used. The resulting system can then be compressed into a single equation equivalent to (1.5.7).

Also, we can permit the use of arbitrary Diophantine terms as arguments of Diophantine relations in generalized Diophantine representations. In fact, if $\mathcal{R}$ is a Diophantine relation, then the assertion

$$\mathcal{R}(\mathfrak{t}_1, \ldots, \mathfrak{t}_n) \tag{1.5.9}$$

is equivalent to the assertion

$$\exists t_1 \ldots t_n \, [\mathcal{R}(t_1, \ldots, t_n) \,\&\, t_1 = \mathfrak{t}_1 \,\&\, \ldots \,\&\, t_n = \mathfrak{t}_n]. \tag{1.5.10}$$

The first conjunctive term in (1.5.10) can be replaced by a copy of the equation from (1.5.3), and the remaining conjunctive terms can be dealt with as with (1.5.7) above.

A Diophantine function need not be defined for all possible values of its arguments. However, in adjoining equation (1.5.8), we impose the condition that F is defined for $\beta_1, \ldots, \beta_n$. That is why, for example, the formula

$$F(a_1, \ldots, a_n) = 0 \vee F(a_1, \ldots, a_n) \neq 0 \tag{1.5.11}$$

is not a generalized Diophantine representation for the relation that is identically true unless F is defined everywhere.

Note that the term "generalized Diophantine representation" should not be regarded as designating a well-defined mathematical concept; this is because such a representation reflects not only objective qualities of sets, properties, relations, or functions, but also the particular order in which we consider them. For example, we could first give for the property *is an even number* the Diophantine representation

$$\text{Even}(a) \iff \exists x \, [2x = a] \tag{1.5.12}$$

and then consider for the property *is an odd number* the generalized Diophantine representation

$$\text{Odd}(a) \iff \text{Even}(a+1) \tag{1.5.13}$$

or, vice versa, we could first give for Odd the genuine Diophantine representation

$$\text{Odd}(a) \iff \exists x \, [2x + 1 = a] \tag{1.5.14}$$

and then consider the generalized Diophantine representation

$$\text{Even}(a) \iff \text{Odd}(a+1). \tag{1.5.15}$$

However, we cannot *simultaneously* consider both (1.5.13) and (1.5.15) as generalized Diophantine representations.

1.6 Some simple examples of Diophantine sets, properties, relations, and functions

Representation (1.5.1) proves that the relation $\neq$ is Diophantine. Similar proofs can be given for the inequalities $\leq$ and $<$:

$$a \leq b \iff \exists x\,[a+x=b], \tag{1.6.1}$$

$$a < b \iff \exists x\,[a+x+1=b]. \tag{1.6.2}$$

The relation of divisibility is also Diophantine:

$$a \mid b \iff \exists x\,[ax=b]. \tag{1.6.3}$$

This enables us to give the following generalized Diophantine representation for the function *remainder on dividing b by c*, written $\text{rem}(b,c)$:

$$a = \text{rem}(b,c) \iff a < c \,\&\, c \mid b-a. \tag{1.6.4}$$

In Section 2.2 we shall need the function $\text{arem}(b,c)$ whose value is the least absolute value $|\chi|$ among all numbers χ congruent to b with respect to the modulus c:

$$\text{arem}(b,c) \equiv \pm b \pmod{c} \;\;\&\;\; 0 \leq \text{arem}(b,c) \leq c/2. \tag{1.6.5}$$

This function is also Diophantine:

$$a = \text{arem}(b,c) \iff 2a \leq c \,\&\, [c \mid (b-a) \vee c \mid (b+a)]\,. \tag{1.6.6}$$

We shall also need the fact that the relation of non-divisibility is Diophantine:

$$a \nmid b \iff \text{rem}(b,a) > 0. \tag{1.6.7}$$

Note that the relation $\nmid$ introduced in this way is not the negation of the relation of divisibility $\mid$ because according to (1.6.3) and (1.6.7) both relations are false when $a=0$ and $b>0$. This feature could easily be avoided, but in fact we shall use the relation $\nmid$ only in cases when it is certain that the first argument is not zero.

Having proved that rem is Diophantine, we can give a generalized Diophantine representation for the function *integer part of* b/c, written $b \operatorname{div} c$, namely,

$$a = b \operatorname{div} c \iff ac + \operatorname{rem}(b, c) = b \tag{1.6.8}$$

and a generalized Diophantine representation of congruence with respect to a positive modulus:

$$a \equiv b \pmod{c} \iff \operatorname{rem}(a, c) = \operatorname{rem}(b, c). \tag{1.6.9}$$

It is known that $\gcd(b, c)$, *the greatest common divisor of the positive integers* b *and* c, can be represented in the form $bx - cy$, and clearly, every number of this form is divisible by $\gcd(b, c)$. Thus, we have the following generalized Diophantine representation for gcd:

$$a = \gcd(b, c) \iff bc > 0 \,\&\, a \mid b \,\&\, a \mid c \,\&\, \exists xy\, [a = bx - cy]. \tag{1.6.10}$$

In turn, gcd being Diophantine allows us to show that the function $\operatorname{lcm}(b, c)$, *the least common multiple of the positive integers* b *and* c, is also:

$$a = \operatorname{lcm}(b, c) \iff bc = a \gcd(b, c). \tag{1.6.11}$$

Exercises

1. In Section 1.3 we established the equivalence of two decision problems about solvability of Diophantine equations in integers and in natural numbers. Show that they are also equivalent to the analogous problem concerning solvability in positive integers.

2. Passing from equation (1.3.2) to system (1.3.3) results in a five-fold increase in the number of unknowns. We can eliminate the unknowns $x_1, \ldots, x_m$ using their explicit expression in terms of $y_{1,1}, \ldots, y_{m,4}$, which results in only a four-fold increase in the number of unknowns. Find another method resulting in an even smaller increase in the number of unknowns.

3. Passing from equation (1.3.5) to equation (1.3.6) doubles the number of unknowns. Show that the problem of the solvability of (1.3.5) in integers can be reduced to the problem of the solvability in natural numbers of some Diophantine equation with m unknowns.

4. Equation (1.4.7) is equivalent to equation (1.4.8) only under our convention that the parameter a and the unknowns are natural numbers. Show that a set of natural numbers is Diophantine if and only if it is the set of all natural numbers assumed by some polynomial with integer coefficients for arbitrary *integral* values of the variables.

5. It is easy to see that every finite set is Diophantine but that no finite set containing more than one element can be the set of *all* values assumed by some polynomial with integer coefficients for natural number (or arbitrary integer) values of the variables. For a less trivial example, show that no set that contains at least two elements and consists only of prime numbers can be represented in this form either, even if one considers the numbers -2, -3, -5, ... to be prime. (It will be shown in Section 3.4 that the set of all prime numbers is Diophantine).

6. According to the previous exercise, there must be, among the values assumed by the polynomial from the left-hand side of (1.4.8), some negative integers in addition to the elements of the set defined by equation (1.4.7). Show that for every Diophantine set $\mathfrak{M}$ of natural numbers there is a polynomial with integer coefficients whose range (i.e., the set of *all* values that it assumes) is the union of the set $\mathfrak{M}$ and the set of all negative integers (the variables can be either natural numbers or arbitrary integers).

7. Show that for every Diophantine set $\mathfrak{M}$ of natural numbers there are polynomials P and Q with integer coefficients such that the set $\mathfrak{M}$ is exactly the set of *all integer* values assumed by the fraction

$$\frac{P(x_0, \ldots, x_m)}{Q(x_0, \ldots, x_m)}$$

for natural number values of the variables.

8. The special form of Diophantine representations of one-dimensional sets from Section 1.4 has different generalizations for the multidimensional case. Show that

(a) every n-dimensional Diophantine set $\mathfrak{M}$ has a representation of the form

$$D(x_1, \ldots, x_m) = A(a_1, \ldots, a_n)$$

where A and D are polynomials with integer coefficients;
(b) for every such set $\mathfrak{M}$ there are also polynomials D_1, ..., D_n with integer

coefficients such that $\langle a_1, \ldots, a_n \rangle \in \mathfrak{M}$ if and only if the system

$$D_1(x_1, \ldots, x_m) = a_1$$
$$\vdots$$
$$D_n(x_1, \ldots, x_m) = a_n$$

has a solution.

9. Show that every Diophantine function has a representation of the form

$$a = F(b_1, \ldots, b_n) \iff \exists x_0 \ldots x_m \, [a = E(b_1, \ldots, b_n, x_0, \ldots, x_m)]$$

where E is a polynomial with integer coefficients.

10. Show that the following sets are Diophantine:

(a) the set of all composite numbers;
(b) the set of all numbers that are not powers of 2;
(c) the set of all numbers that are not perfect squares.

11. For a less trivial example, show that the relation *a is not a power of b* is Diophantine.

12. Show that for every positive algebraic number χ the two-place relation *p/q is a convergent of the continued fraction for χ* is Diophantine.

13. Let a *trigonometric Diophantine equation* be an equation of the form

$$R\left(x_1, \ldots, x_m, T_1\left(\frac{P_1(x_1, \ldots, x_m)}{Q_1(x_1, \ldots, x_m)}\pi\right), \ldots, T_l\left(\frac{P_l(x_1, \ldots, x_m)}{Q_l(x_1, \ldots, x_m)}\pi\right)\right) = 0$$

where $P_1, \ldots, P_l, Q_1, \ldots, Q_l, R$ are polynomials with integer coefficients and $T_1, \ldots, T_l$ are trigonometric functions. Show that the problem of testing for the existence of a solution of a trigonometric Diophantine equation is reducible to Hilbert's Tenth Problem.

14. Show that if $R(x_1, \ldots, x_m)$ is a polynomial of degree k with *real* coefficients that assumes only integral values when $x_1, \ldots, x_m$ are natural numbers, then

$$R(x_1, \ldots, x_m) = P\left(\binom{x_1}{1}, \ldots, \binom{x_1}{k}, \ldots, \binom{x_m}{1}, \ldots, \binom{x_m}{k}\right)$$

for some polynomial P with integer coefficients.

15. As was mentioned in Section 1.4, in the *general* case the complement of a Diophantine set need not be Diophantine. For the *special* case of a Diophantine set defined by an equation with one unknown, show that its complement is Diophantine.

Unsolved problems

1. Can every Diophantine set be defined by an equation of degree 3 with respect to all its variables? With respect to all unknowns?

2. Is Hilbert's Tenth Problem reducible to its restriction to equations of degree 3?

3. Is the restriction of Hilbert's Tenth Problem to equations of degree 3 undecidable?

Commentary

All of Diophantus's extant writings are available in English (Heath [1910]) and Russian (Diophantus [1974]) translations with commentaries.

Hilbert's Tenth Problem is the only decision problem among the twenty-three problems that Hilbert listed.

Number theorists did not show much interest in the inverse problem of finding for a set, property, or relation a Diophantine representation or proving the impossibility of such a representation. Problems of that kind are more typical for investigations in mathematical logic. From "The Autobiography of Julia Robinson" (see Reid [1986]), we know that in 1948 Alfred Tarski posed the problem of proving that the set of all powers of 2 is not Diophantine. For Julia Robinson [1952], this problem was the starting point for a systematic investigation to determine whether various particular sets are Diophantine. In this same period, Martin Davis [1953] analyzed the class of Diophantine sets as a whole. In particular, he showed that this class is closed under union and intersection but not under complementation. Also, he established the equivalence of Hilbert's Tenth Problem to its counterpart for solutions in natural numbers. Another paper from that period of pioneering investigations on Hilbert's Tenth Problem is Myhill [1953].

The special forms of Diophantine representations of one-dimensional sets from Section 1.4 and Exercises 1.4, 1.6, and 1.7 were proposed by Hilary Putnam [1960].

The general Diophantine equation was reduced to one of degree 4 by Thoralf Skolem [1934]. Reduction of a general Diophantine equation to a system of Diophantine equations of degree 2 having a special form was considered by Britton [1979].

The problem of the algorithmic unsolvability of equations of degree 3 was stated in Davis, Matiyasevich, and Robinson [1976]. The three unsolved problems stated above are, of course, closely related to one another. Namely, a positive answer to the first implies a positive answer to the second, which in turn implies a positive answer to the third. However, the three problems are not equivalent; *a priori*, the answer to the first problem could turn out to be negative while the answer to the third was positive. For the counterparts of these three problems for the case of equations of degree 2, negative answers follow from Siegel [1972].

2 Exponentiation Is Diophantine

The goal of this chapter is to show that the function b^c is a Diophantine function of two arguments. The proof is rather technical (and Sections 2.1–2.4 could be skipped at the first reading), but it opens a straightforward path to proving that many other interesting and important functions and relations are Diophantine. In particular, in Section 3.4 we show that the set of all prime numbers is Diophantine.

2.1 Special second-order recurrent sequences

As was just stated, the goal of this chapter is to show that the function b^c is Diophantine or, equivalently, to show that the set of triples

$$\{ \langle a, b, c \rangle \mid a = b^c \} \tag{2.1.1}$$

is Diophantine. Clearly, this would imply that the set of pairs

$$\{ \langle a, b \rangle \mid \exists n \, [a = b^n] \} \tag{2.1.2}$$

is Diophantine. The converse implication is also valid (see Exercise 2.2), but it is not easy even to establish that (2.1.2) is Diophantine.

The set of all powers of a given number b may be viewed as the set of all members of the first-order recurrent sequence

$$\beta_b(0) = 1, \qquad \beta_b(n+1) = b\beta_b(n). \tag{2.1.3}$$

In our proof an essential role will be played by the second-order recurrent sequence

$$\alpha_b(0) = 0, \qquad \alpha_b(1) = 1, \qquad \alpha_b(n+2) = b\alpha_b(n+1) - \alpha_b(n), \tag{2.1.4}$$

where $b \geq 2$. In this section we take the first step by showing that the set of pairs

$$\{ \langle a, b \rangle \mid b \geq 2 \,\&\, \exists n \, [a = \alpha_b(n)] \} \tag{2.1.5}$$

is Diophantine. Strange as it may seem, it is much easier to prove this than it is to show that the set (2.1.2), which is so much more natural and commonplace, is Diophantine.

The second-order relation (2.1.4) can be rewritten as a first-order relation among the matrices

$$\mathrm{A}_b(n) = \begin{pmatrix} \alpha_b(n+1) & -\alpha_b(n) \\ \alpha_b(n) & -\alpha_b(n-1) \end{pmatrix}, \tag{2.1.6}$$

taking $\alpha_b(-1) = -1$. Namely,

$$\mathrm{A}_b(0) = \mathrm{E}, \qquad \mathrm{A}_b(n+1) = \mathrm{A}_b(n)\Xi_b, \tag{2.1.7}$$

where

$$\mathrm{E} = \begin{pmatrix} 1 & 0 \\ 0 & 1 \end{pmatrix}, \qquad \Xi_b = \begin{pmatrix} b & -1 \\ 1 & 0 \end{pmatrix}. \tag{2.1.8}$$

This implies that

$$\mathrm{A}_b(n) = \Xi_b^n, \tag{2.1.9}$$

and hence

$$\det(\mathrm{A}_b(n)) = 1, \tag{2.1.10}$$

i.e.,

$$\begin{aligned} \alpha_b^2(n) - \alpha_b(n+1)\alpha_b(n-1) &= \alpha_b^2(n+1) - b\alpha_b(n+1)\alpha_b(n) + \alpha_b^2(n) \\ &= \alpha_b^2(n-1) - b\alpha_b(n-1)\alpha_b(n) + \alpha_b^2(n) = 1. \end{aligned} \tag{2.1.11}$$

It turns out that equation (2.1.11) characterizes the sequence (2.1.4) in the following sense: *if*

$$x^2 - bxy + y^2 = 1, \tag{2.1.12}$$

then either

$$x = \alpha_b(m+1), \qquad y = \alpha_b(m) \tag{2.1.13}$$

or

$$x = \alpha_b(m), \qquad y = \alpha_b(m+1) \tag{2.1.14}$$

for some m. In order to distinguish which of the two cases, (2.1.13) or (2.1.14), holds, it is sufficient to note that (2.1.4) implies by induction that

$$0 = \alpha_b(0) < \alpha_b(1) < \cdots < \alpha_b(n) < \alpha_b(n+1) < \cdots \tag{2.1.15}$$

We now show that *equation* (2.1.12) *together with the inequality*

$$y < x \tag{2.1.16}$$

implies the existence of some m for which (2.1.13) *holds.* The proof will proceed by induction on y. If $y = 0$, then clearly $x = 1$; i.e., (2.1.13) holds with $m = 0$. If $y > 0$, then (2.1.12) and (2.1.16) imply that

$$x = by + \frac{1 - y^2}{x} \leq by, \tag{2.1.17}$$

$$x = by + \frac{1}{x} - \frac{y^2}{x} > by - y. \tag{2.1.18}$$

Let $x_1 = y$ and $y_1 = by - x$. Then

$$\begin{aligned} x_1^2 - bx_1y_1 + y_1^2 &= y^2 - by(by - x) + (by - x)^2 \\ &= x^2 - bxy + y^2 \\ &= 1. \end{aligned} \tag{2.1.19}$$

By (2.1.18), $y_1 < x_1$, and by the induction hypothesis,

$$x_1 = \alpha_b(m_1 + 1), \qquad y_1 = \alpha_b(m_1) \tag{2.1.20}$$

for some m_1. Hence, for $m = m_1 + 1$,

$$x = bx_1 - y_1 = \alpha_b(m + 1), \qquad y = x_1 = \alpha_b(m). \tag{2.1.21}$$

Thus, we have proved that the set (2.1.5) is defined by the formula

$$b \geq 2 \,\&\, \exists x \left[x^2 - abx + a^2 = 1\right]. \tag{2.1.22}$$

2.2 The special recurrent sequences are Diophantine (basic ideas)

Our first goal will be to show that the set of triples

$$\{ \langle a, b, c \rangle \mid b \geq 4 \,\&\, a = \alpha_b(c) \} \tag{2.2.1}$$

is Diophantine. In this section we outline the underlying ideas, while the formal proof will be given in the next section.

It is convenient to consider the set (2.2.1) as the union of the terms of the sequences

$$\langle \alpha_b(0), b, 0 \rangle, \ldots, \langle \alpha_b(n), b, n \rangle, \ldots \tag{2.2.2}$$

for $b = 4, 5, \ldots$. Using induction on definition (2.1.4), it is easy to derive that

$$\alpha_2(n) = n. \tag{2.2.3}$$

Hence, for $b = 2$ the sequence (2.2.2) is very simple:

$$\langle 0, 2, 0 \rangle, \ldots, \langle n, 2, n \rangle, \ldots \tag{2.2.4}$$

However, we are concerned with the case $b \geq 4$, so $\alpha_b(n)$ cannot be defined by a simple equation like (2.2.3). Nevertheless, for $b > 2$ there is a weak analog of (2.2.3); namely, it follows by induction from (2.1.4) that

$$\alpha_{b_1}(n) \equiv \alpha_{b_2}(n) \pmod q \tag{2.2.5}$$

provided that

$$b_1 \equiv b_2 \pmod q. \tag{2.2.6}$$

Hence, in particular,

$$\alpha_b(n) \equiv \alpha_2(n) = n \pmod{b-2}, \tag{2.2.7}$$

so that the first $b-2$ members of the sequence (2.2.2) coincide with the first $b-2$ members of the sequence

$$\langle \alpha_b(0), b, \mathrm{rem}(\alpha_b(0), b-2) \rangle, \ldots, \langle \alpha_b(n), b, \mathrm{rem}(\alpha_b(n), b-2) \rangle, \ldots \tag{2.2.8}$$

The "advantage" of the sequence (2.2.8) (as compared with (2.2.2)) consists in the fact that here n enters only as an argument of α. Together with the facts that the set (2.1.5) and the function rem are Diophantine, this implies that the set of all triples from (2.2.8) is also Diophantine. The "disadvantage" consists in the fact that only finite initial segments of the sequences (2.2.8) and (2.2.2) are equal.

Using (2.2.5), we can construct another sequence that has the same "advantage" but avoids the "disadvantage." Namely, let

$$w \equiv b \pmod v, \tag{2.2.9}$$

$$w \equiv 2 \pmod u, \tag{2.2.10}$$

$$v > 2\alpha_b(k), \tag{2.2.11}$$

$$u > 2k. \tag{2.2.12}$$

Then, as we shall see, the first k members of the sequence (2.2.2) coincide with the first k members of the sequence

$$\langle \operatorname{arem}(\alpha_w(0), v), b, \operatorname{arem}(\alpha_w(0), u)\rangle, \\ \ldots, \langle \operatorname{arem}(\alpha_w(n), v), b, \operatorname{arem}(\alpha_w(n), u\rangle, \ldots \quad (2.2.13)$$

(The function rem is replaced here by the function arem that was introduced in Section 1.6; the role of this substitution will become clear later.)

Now, the union of all the sequences of the form (2.2.13) with u, v, and w satisfying conditions (2.2.9) and (2.2.10) certainly contains all the triples from the sequence (2.2.2); however, this union may also contain some additional triples. To eliminate these additional triples, we begin by imposing on u and v, besides (2.2.9) and (2.2.10), some further conditions, and, moreover, we exclude from (2.2.13) those triples that do not satisfy the inequality

$$2 \operatorname{arem}\big(\alpha_w(n), v\big) < u. \quad (2.2.14)$$

In order to understand the nature of these new conditions on v and u, we note that the recurrent relations (2.1.4) imply that for any positive v the sequence

$$\alpha_b(0), \ldots, \alpha_b(n), \ldots \quad (2.2.15)$$

is purely periodic modulo v. For our special choice of v we will be able to determine the length of the period and its structure. Namely, let

$$v = \alpha_b(m+1) - \alpha_b(m-1); \quad (2.2.16)$$

then

$$\alpha_b(m+1) \equiv \alpha_b(m-1) \pmod{v}. \quad (2.2.17)$$

The recurrent relation (2.1.4) can be rewritten as

$$\alpha_b(n-2) = b\alpha_b(n-1) - \alpha_b(n), \quad (2.2.18)$$

and hence

$$\begin{aligned} \alpha_b(m+2) &= b\alpha_b(m+1) - \alpha_b(m) \\ &\equiv b\alpha_b(m-1) - \alpha_b(m) \pmod{v} \\ &= \alpha_b(m-2), \end{aligned} \quad (2.2.19)$$

$$\begin{aligned} \alpha_b(m+3) &\equiv \alpha_b(m-3) \pmod{v}, \\ &\vdots \\ \alpha_b(2m-1) &\equiv \alpha_b(1) \pmod{v}, \\ \alpha_b(2m) &\equiv \alpha_b(0) \pmod{v}. \end{aligned} \tag{2.2.20}$$

Furthermore, we have

$$\alpha_b(2m) \equiv \alpha_b(0) = 0 = -\alpha_b(0) \pmod{v}, \tag{2.2.21}$$

$$\alpha_b(2m+1) = b\alpha_b(2m) - \alpha_b(2m-1) \equiv -\alpha_b(1) \pmod{v} \tag{2.2.22}$$

and hence

$$\alpha_b(2m+n) \equiv -\alpha_b(n) \pmod{v}. \tag{2.2.23}$$

Thus, for our choice of v, the sequence (2.2.15) modulo v has the following period of $4m$ terms:

$$\begin{array}{ccccccc} 0, & 1, \dots, & \alpha_b(m-1), & \alpha_b(m), & \alpha_b(m-1), \dots, & 1, \\ 0, & -1, \dots, & -\alpha_b(m-1), & -\alpha_b(m), & -\alpha_b(m-1), \dots, & -1. \end{array} \tag{2.2.24}$$

According to (2.2.9), this is also the period modulo v of the sequence

$$\alpha_w(0), \dots, \alpha_w(n), \dots \tag{2.2.25}$$

Correspondingly, the sequence

$$\text{arem}(\alpha_w(0), v), \dots, \text{arem}(\alpha_w(n), v), \dots \tag{2.2.26}$$

has the period of $2m$ terms

$$0, 1, \dots, \alpha_b(m-1), \alpha_b(m), \alpha_b(m-1), \dots, 1, \tag{2.2.27}$$

because according to (2.1.15)

$$\begin{aligned} v &= \alpha_b(m+1) - \alpha_b(m-1) \\ &= b\alpha_b(m) - 2\alpha_b(m-1) \\ &\geq 2\alpha_b(m) \end{aligned}$$

for $b \geq 4$.

According to (2.2.10) and (2.2.7), the sequence (2.2.25) modulo u has the period of u terms

$$0, 1, \dots, u-1. \tag{2.2.28}$$

Now we impose on u the very important condition

$$u \mid m. \tag{2.2.29}$$

This implies that the length of the period of the sequence (2.2.26) is a multiple of the length of the period of the sequence

$$\operatorname{arem}\big(\alpha_w(0), u\big), \ldots, \operatorname{arem}\big(\alpha_w(n), u\big), \ldots \tag{2.2.30}$$

and hence that the sequence (2.2.13) has an almost symmetrical period of length $2m$. Thus all the "extra" triples in (2.2.13) should appear among the first $m+1$ members of this sequence. For these initial triples, condition (2.2.14) can be rewritten as

$$2\alpha_b(n) < u, \tag{2.2.31}$$

and therefore

$$2n < u, \tag{2.2.32}$$

because according to (2.1.15)

$$n \leq \alpha_b(n). \tag{2.2.33}$$

Now, (2.2.32) implies that

$$\operatorname{arem}\big(\alpha_b(n), u\big) = \operatorname{arem}(n, u) = n; \tag{2.2.34}$$

thus, condition (2.2.14) indeed eliminates all the "extra" triples.

In trying to implement the plan described above, we encounter the following difficulty: how can we transform the pair of conditions (2.2.16) and (2.2.29) into Diophantine equations without first proving that α is a Diophantine function? To overcome this difficulty we shall employ the following property of α:

$$\alpha_b^2(k) \mid \alpha_b(m) \Longrightarrow \alpha_b(k) \mid m. \tag{2.2.35}$$

We shall put

$$u = \alpha_b(k) \tag{2.2.36}$$

and replace (2.2.29) by the stronger condition

$$u^2 \mid \alpha_b(m). \tag{2.2.37}$$

2.3 The special recurrent sequences are Diophantine (proof)

We begin by proving the implication (2.2.35). Let b, k, and m satisfy

$$\alpha_b^2(k) \mid \alpha_b(m). \tag{2.3.1}$$

Recall that $\alpha_b(k)$ and $\alpha_b(m)$ are elements of the matrices $\mathrm{A}_b(k)$ and $\mathrm{A}_b(m)$ defined by (2.1.6) and satisfying (2.1.9). Let

$$m = n + kl, \qquad 0 \leq n < k. \tag{2.3.2}$$

We have

$$\begin{aligned}\begin{pmatrix}\alpha_b(m+1) & -\alpha_b(m)\\ \alpha_b(m) & -\alpha_b(m-1)\end{pmatrix} &= \mathrm{A}_b(m)\\ &= \Xi_b^m\\ &= \Xi_b^{n+kl}\\ &= \Xi_b^n(\Xi_b^k)^l\\ &= \mathrm{A}_b(n)\mathrm{A}_b^l(k)\\ &= \begin{pmatrix}\alpha_b(n+1) & -\alpha_b(n)\\ \alpha_b(n) & -\alpha_b(n-1)\end{pmatrix}\begin{pmatrix}\alpha_b(k+1) & -\alpha_b(k)\\ \alpha_b(k) & -\alpha_b(k-1)\end{pmatrix}^l.\end{aligned} \tag{2.3.3}$$

Passing to a congruence modulo $\alpha_b(k)$, we obtain

$$\begin{aligned}&\begin{pmatrix}\alpha_b(m+1) & -\alpha_b(m)\\ \alpha_b(m) & -\alpha_b(m-1)\end{pmatrix} \equiv\\ &\qquad\begin{pmatrix}\alpha_b(n+1) & -\alpha_b(n)\\ \alpha_b(n) & -\alpha_b(n-1)\end{pmatrix}\begin{pmatrix}\alpha_b(k+1) & 0\\ 0 & -\alpha_b(k-1)\end{pmatrix}^l \pmod{\alpha_b(k)},\end{aligned} \tag{2.3.4}$$

and hence

$$\alpha_b(m) \equiv \alpha_b(n)\alpha_b^l(k+1) \pmod{\alpha_b(k)}. \tag{2.3.5}$$

By (2.1.11), $\alpha_b(k)$ and $\alpha_b(k+1)$ are coprime; thus, (2.3.1) and (2.3.5) imply that

$$\alpha_b(k) \mid \alpha_b(n). \tag{2.3.6}$$

Now it follows from (2.3.2) and (2.1.15) that $\alpha_b(n) < \alpha_b(k)$, so that (2.3.6) is possible only if $n = 0$, i.e., if $m = kl$. Furthermore, we have:

$$\begin{aligned}\mathrm{A}_b(m) &= \mathrm{A}_b^l(k)\\ &= [\alpha_b(k)\Xi_b - \alpha_b(k-1)\mathrm{E}]^l\\ &= \sum_{i=0}^{l}(-1)^{l-i}\binom{l}{i}\alpha_b^i(k)\alpha_b^{l-i}(k-1)\Xi_b^i.\end{aligned} \tag{2.3.7}$$

Passing from the equality to a congruence modulo $\alpha_b^2(k)$, we can omit all the summands except the first two:

$$\begin{aligned}\mathrm{A}_b(m) &= \begin{pmatrix}\alpha_b(m+1) & -\alpha_b(m)\\ \alpha_b(m) & -\alpha_b(m-1)\end{pmatrix}\\ &\equiv (-1)^l\alpha_b^l(k-1)\mathrm{E} \;+\; (-1)^{l-1}l\alpha_b(k)\alpha_b^{l-1}(k-1)\Xi_b \pmod{\alpha_b^2(k)},\end{aligned} \tag{2.3.8}$$

whence

$$\alpha_b(m) \equiv (-1)^{l-1}l\alpha_b(k)\alpha_b^{l-1}(k-1) \pmod{\alpha_b^2(k)}. \tag{2.3.9}$$

Together with (2.3.1) this implies that

$$\alpha_b(k) \mid l\alpha_b^{l-1}(k-1), \tag{2.3.10}$$

and because by (2.1.11) $\alpha_b(k)$ and $\alpha_b(k-1)$ are coprime,

$$\alpha_b(k) \mid l. \tag{2.3.11}$$

The implication (2.2.35) is proved.

Now we can exhibit a system of Diophantine conditions that is solvable if and only if the triple $\langle a, b, c\rangle$ belongs to the set (2.2.1):

$$b \geq 4, \tag{2.3.12}$$

$$u^2 - but + t^2 = 1, \tag{2.3.13}$$

$$s^2 - bsr + r^2 = 1, \tag{2.3.14}$$

$$r < s, \tag{2.3.15}$$

$$u^2 \mid s, \tag{2.3.16}$$

$$v = bs - 2r, \tag{2.3.17}$$

$$v \mid w - b, \tag{2.3.18}$$

$$u \mid w - 2, \tag{2.3.19}$$
$$w > 2, \tag{2.3.20}$$
$$x^2 - wxy + y^2 = 1, \tag{2.3.21}$$
$$2a < u, \tag{2.3.22}$$
$$a = \operatorname{arem}(x, v), \tag{2.3.23}$$
$$c = \operatorname{arem}(x, u). \tag{2.3.24}$$

We first prove that if the conditions (2.3.12)–(2.3.24) are satisfied, then

$$a = \alpha_b(c). \tag{2.3.25}$$

It was shown in Section 2.1 that (2.3.12) and (2.3.13) imply that for some k,

$$u = \alpha_b(k). \tag{2.3.26}$$

Likewise, (2.3.12), (2.3.14), and (2.3.15) imply that for some positive m,

$$s = \alpha_b(m), \qquad r = \alpha_b(m-1). \tag{2.3.27}$$

By (2.2.35), it follows from (2.3.16), (2.3.26), and (2.3.27) that

$$u \mid m. \tag{2.3.28}$$

By (2.1.4), it follows from (2.3.17) and (2.3.27) that

$$v = \alpha_b(m+1) - \alpha_b(m-1). \tag{2.3.29}$$

Furthermore, it follows from (2.3.20) and (2.3.21) that for some n,

$$x = \alpha_w(n). \tag{2.3.30}$$

From this, (2.3.18) and (2.3.19), and (2.2.5)–(2.2.7), we have that

$$x \equiv \alpha_b(n) \pmod{v}, \tag{2.3.31}$$
$$x \equiv n \pmod{u}. \tag{2.3.32}$$

Let

$$n = 2lm \pm j, \tag{2.3.33}$$

where

$$j \leq m. \tag{2.3.34}$$

Using the matrix representation once again, we have:

$$\begin{aligned} \mathrm{A}_b(n) &= \Xi_b^n \\ &= \Xi_b^{2lm \pm j} \\ &= \left[[\Xi_b^m]^2\right]^l \Xi_b^{\pm j} \\ &= \left[[\mathrm{A}_b(m)]^2\right]^l [\mathrm{A}_b(j)]^{\pm 1}, \end{aligned} \tag{2.3.35}$$

$$\begin{aligned} \mathrm{A}_b(m) &= \begin{pmatrix} \alpha_b(m+1) & -\alpha_b(m) \\ \alpha_b(m) & -\alpha_b(m-1) \end{pmatrix} \\ &\equiv - \begin{pmatrix} -\alpha_b(m-1) & \alpha_b(m) \\ -\alpha_b(m) & \alpha_b(m+1) \end{pmatrix} \pmod{v} \\ &= -[\mathrm{A}_b(m)]^{-1}, \end{aligned} \tag{2.3.36}$$

$$[\mathrm{A}_b(m)]^2 \equiv -\mathrm{E} \pmod{v} \tag{2.3.37}$$

$$\mathrm{A}_b(n) \equiv \pm[\mathrm{A}_b(j)]^{\pm 1} \pmod{v}. \tag{2.3.38}$$

(In this last formula all four combinations of the signs "+" and "−" are possible.) Passing from the matrix congruence (2.3.38) to element-wise congruence, we have that

$$x \equiv \alpha_b(n) \equiv \pm\alpha_b(j) \pmod{v}. \tag{2.3.39}$$

By (2.1.15), it follows from (2.3.34) that

$$2\alpha_b(j) \leq 2\alpha_b(m) \leq (b-2)\alpha_b(m) < b\alpha_b(m) - 2\alpha_b(m-1) = v, \tag{2.3.40}$$

and hence

$$a = \mathrm{arem}(x, v) = \mathrm{arem}(\alpha_b(n), v) = \alpha_b(j). \tag{2.3.41}$$

From this and (2.3.22), using (2.2.33), we have that

$$2j \leq 2\alpha_b(j) = 2a < u. \tag{2.3.42}$$

Finally, from (2.3.28), (2.3.31), (2.3.33), and (2.3.42) we obtain

$$c = \mathrm{arem}(x, u) = \mathrm{arem}(n, u) = j, \tag{2.3.43}$$

which together with (2.3.41) gives the desired equality (2.3.25).

Now we are going to prove the converse; i.e., we will show that if the numbers a, b, and c satisfy (2.3.12) and (2.3.25), then there are numbers s, r, u, t, v, w satisfying (2.3.13)–(2.3.24). The above considerations indicate how these numbers are to be chosen.

We begin by choosing u according to (2.3.26), selecting a and k so that the inequality (2.3.22) holds and u is odd. We are able to do this because by (2.1.15) the sequence $\alpha_b(0)$, $\alpha_b(1)$, ... increases monotonically and by (2.1.11) at least one of any two consecutive terms of the sequence is odd. Let

$$t = \alpha_b(k+1); \tag{2.3.44}$$

then using (2.1.11), equation (2.3.13) holds.

We choose r and s as in (2.3.27), with

$$m = uk; \tag{2.3.45}$$

then by (2.1.11) and (2.1.15), equation (2.3.14) and inequality (2.3.15) both hold. Using (2.3.9),

$$s = \alpha_b(uk) \equiv (-1)^{u-1} u \alpha_b(k) \alpha_b^{u-1}(k-1) \pmod{u^2}; \tag{2.3.46}$$

hence condition (2.3.16) is also valid.

We can find v satisfying (2.3.17) because using (2.1.15)

$$bs - 2r \geq 4\alpha_b(m) - 2\alpha_b(m-1) > 2\alpha_b(m). \tag{2.3.47}$$

We now verify that u and v are coprime. Suppose that $d|u$ and $d|v$; then by (2.3.16) $d \mid s$ and by (2.3.17) $d \mid 2r$. However, by our choice of u, d is odd; hence $d \mid r$ and by (2.3.14) $d \mid 1$. Thus by the Chinese Remainder Theorem (see the Appendix) we can find w satisfying (2.3.18), (2.3.19), and (2.3.20).

Finally, let

$$x = \alpha_w(c), \qquad y = \alpha_w(c+1); \tag{2.3.48}$$

then by (2.1.15), equation (2.3.21) holds.

Using (2.2.5) it follows from (2.3.25), (2.3.48), and (2.3.18) that

$$x = \alpha_w(c) \equiv \alpha_b(c) = a \pmod{v}. \tag{2.3.49}$$

From (2.3.17), (2.3.25), and (2.3.47) it follows that

$$v > 2a, \tag{2.3.50}$$

and hence (2.3.49) implies (2.3.23).

By (2.1.7) it follows from (2.3.48) that

$$x \equiv c \pmod{w-2}, \tag{2.3.51}$$

which together with (2.3.19) gives the congruence

$$x \equiv c \pmod{u}. \tag{2.3.52}$$

By (2.2.33) it follows from (2.3.25) and (2.3.22) that

$$2c \le 2\alpha_b(c) = 2a < u, \tag{2.3.53}$$

which together with (2.3.52) implies (2.3.24).

All of the conditions (2.3.12)–(2.3.24) are Diophantine; thus, we have established that the set (2.2.1) is Diophantine.

2.4 Exponentiation is Diophantine

To begin with, we need to specify the value of 0^0. For a number of different reasons, it is convenient to make the definition $0^0 = 1$.

The recurrent relation (2.1.4) is close to (2.1.3) for large values of b, and $\alpha_b(n)$ grows approximately like $\beta_b(n) = b^n$. More precisely, it is easy to prove by induction that

$$(b-1)^n \le \alpha_b(n+1) \le b^n. \tag{2.4.1}$$

For a fixed n the relative error goes to 1 when $b \to \infty$, but we need a good approximation to b^c for a fixed b. That is why we introduce a new variable x with a large value. Later we'll verify that

$$b^c = \lim_{x\to\infty} \frac{\alpha_{bx+4}(c+1)}{\alpha_x(c+1)}. \tag{2.4.2}$$

Moreover, this relation holds for all values of b and c, including the case $b = 0$. Our language of Diophantine equations does not contain the operation lim, but in this case it can be replaced by the function div. Namely, (2.4.1) implies that

$$\frac{\alpha_{bx+4}(c+1)}{\alpha_x(c+1)} \ge \frac{(bx+3)^c}{x^c} \ge b^c \tag{2.4.3}$$

for large x and hence that

$$b^c = \alpha_{bx+4}(c+1) \operatorname{div} \alpha_x(c+1). \tag{2.4.4}$$

In order to determine when the value of x is sufficiently large, we estimate the left-hand side of (2.4.3) from above. We have to treat the cases $b = 0$ and $b > 0$ separately. For $b = c = 0$

$$\frac{\alpha_{bx+4}(c+1)}{\alpha_x(c+1)} = 1; \tag{2.4.5}$$

for $b = 0,\ \ c > 0,\ \ x > 4$,

$$\frac{\alpha_{bx+4}(c+1)}{\alpha_x(c+1)} < \frac{4^c}{(x-1)^c} \leq 1; \tag{2.4.6}$$

for $b > 0,\ \ x > 16c$,

$$\begin{aligned}
\frac{\alpha_{bx+4}(c+1)}{\alpha_x(c+1)} &\leq \frac{(bx+4)^c}{(x-1)^c} \\
&\leq \frac{\left(1+\frac{4}{x}\right)^c}{\left(1-\frac{1}{x}\right)^c} b^c \\
&\leq \frac{b^c}{\left(1-\frac{1}{x}\right)^c\left(1-\frac{4}{x}\right)^c} \\
&\leq \frac{b^c}{\left(1-\frac{4}{x}\right)^{2c}} \\
&\leq \frac{b^c}{1-\frac{8c}{x}} \\
&\leq b^c\left(1+\frac{16c}{x}\right).
\end{aligned} \tag{2.4.7}$$

Thus (2.4.4) becomes valid as soon as

$$x > 16(c+1)(b+1)^c; \tag{2.4.8}$$

for example, we can take

$$x = 16(c+1)\alpha_{b+4}(c+1). \tag{2.4.9}$$

To obtain a Diophantine representation for $a = b^c$, we need to eliminate the function α from (2.4.4) and (2.4.9). For this purpose we can use three copies of the conditions (2.3.13)–(2.3.24) only, because condition (2.3.12) will be fulfilled automatically.

2.5 Exponential Diophantine equations

In what follows, an important role will be played by *exponential Diophantine equations.* These are equations of the form

$$E_1(x_1, \ldots, x_m) = E_2(x_1, \ldots, x_m), \tag{2.5.1}$$

where E_1 and E_2 are expressions constructed from variables and particular natural numbers using addition, multiplication, and exponentiation.

We do not allow the use of subtraction in exponential Diophantine equations in order to remain within the set of natural numbers and thus avoid such problems as specifying the value of

$$(x-y)^{2^{2^{x-y}}} \tag{2.5.2}$$

when $x = 1$, and $y = 3$. However, every now and then we shall use the sign "$-$" when writing down exponential Diophantine equations, but only in cases when this sign could be eliminated by transposing some terms to the other side or by similarly evident transformations.

Bearing this remark in mind, we see that any system of exponential Diophantine equations can be compressed into a single equation in a manner similar to passing from (1.2.1) to (1.2.2).

In analogy with Diophantine equations, one can consider parametric exponential Diophantine equations and introduce *exponential Diophantine representations of sets, properties, relations,* and *functions.* According to Section 1.5, having proved in Section 2.4 that exponentiation is Diophantine, we have a method for transforming any exponential Diophantine equation into an equivalent Diophantine equation with the same parameters, at the cost of an increase in the number of unknowns. Thus the class of sets (properties, relations, functions) having exponential Diophantine representations coincides with the class of Diophantine sets (properties, relations, functions, respectively). Following the convention in Section 1.5, we regard exponential Diophantine representations as being generalized Diophantine representations. However, exponential Diophantine representations may be of interest in themselves, because often they can be more compact than the corresponding genuine Diophantine representations. Also, exponential Diophantine representations may have certain additional properties that we still have been unable to obtain for Diophantine representations.

We can also consider a class of equations that is intermediate between exponential Diophantine equations and genuine Diophantine equations. Namely, a *unary exponential Diophantine equation* is an exponential Diophantine equation in which only constants are raised to powers; i.e., instead of binary exponentials, only unary exponentials such as 2^c, 3^c, ... are used. If only one unary exponential, say 2^c, is used, perhaps several times, then we have a *unary exponential Diophantine equation to the base* 2.

As an example of an exponential Diophantine equation, we consider the famous Fermat equation

$$(p+1)^{s+3} + (q+1)^{s+3} = (r+1)^{s+3}. \tag{2.5.3}$$

It is written here in this form because then *Fermat's Last Theorem* is just the assertion that equation (2.5.3) has no solutions in the unknowns p, q, r, and s. Although (2.5.3) is a Diophantine equation in the unknowns p, q, and r for any fixed value of s, Fermat's Last Theorem, in its original form, is not an individual subproblem of Hilbert's Tenth Problem, because s occurs exponentially in (2.5.3). (Incidentally, Hilbert did not include Fermat's Last Theorem among his "Mathematical Problems.") However, at this point, we are able to construct a specific polynomial $\mathbf{F}$ with integer coefficients such that the equation

$$\mathbf{F}(p,q,r,s,x_1,\ldots,x_m) = 0 \tag{2.5.4}$$

is solvable in $x_1, \ldots, x_m$ if and only if p, q, r, and s satisfy (2.5.3), and hence Fermat's Last Theorem is equivalent to the assertion that (2.5.4) is an unsolvable Diophantine equation in $m+4$ unknowns. Thus, in spite of the fact that we still (1992) don't know whether Fermat's Last Theorem is true or false, we can find an individual subproblem of Hilbert's Tenth Problem to which it is equivalent.

Exercises

In Exercises 2.1–2.3, 2.8–2.10, and in Open Question 2.1, one is supposed to find constructions that are simpler than what would result from a straightforward application of the technique of Chapter 2.

1. Show that if the set (2.1.2) is Diophantine, then so is the set

$$\{\langle a_1, b_1, a_2, b_2\rangle \mid \exists n\,[a_1 = b_1^n \,\&\, a_2 = b_2^n]\}.$$

2. Show that if the set (2.1.2) is Diophantine, then so is the set (2.1.1).

3. For a fixed odd b, give a direct proof that the set

$$\{ a \mid \exists n\, [a = b^n] \}$$

is Diophantine, given that it has an infinite Diophantine subset.

4. Show that for a fixed b, the set (2.1.5) is defined not only by formula (2.1.22) but also by the Pell equation

$$z^2 - \left(\left(\frac{b}{2} \right)^2 - 1 \right) a^2 = 1.$$

5. To prove that exponentiation is Diophantine, we could have used the sequences defined by the relations

$$\gamma_b(0) = 0, \qquad \gamma_b(1) = 1, \qquad \gamma_b(n+2) = b\gamma_b(n+1) + \gamma_b(n),$$

where $b \geq 1$, instead of the sequences (2.1.4). Show that the set

$$\{ \langle a, b \rangle \mid b \geq 1 \,\&\, \exists n\, [a = \gamma_b(n)] \},$$

which is an analog of the set (2.1.5), is Diophantine.

6. In (2.2.1) the inequality $b \geq 4$ is used (instead of $b \geq 2$) for the sake of a slight simplification (where?) of the proof. This limitation was not burdensome in (2.4.4) and (2.4.9) for achieving the main goal of the chapter. For the sake of generality, show that $\alpha_b(c)$ is a Diophantine function of two arguments defined for all $b \geq 2$ and all c.

7. In addition to (2.4.4), there is yet another, less evident, connection between the sequences α and exponentiation, namely

$$b^n \equiv \alpha_x(n+1) + (b - x)\alpha_x(n) \pmod{bx - b^2 - 1}.$$

Prove this and then use it to obtain another Diophantine representation of the set (2.1.1).

8. In (2.4.2) we used the fact (expressed by the inequalities (2.4.1)) that $\alpha_b(n)$ grows almost like b^n. It turns out that these inequalities can be replaced by much weaker ones. Let us say that a binary relation $\mathcal{J}$ *has exponential growth* when the following two conditions hold:

(a) for every u and v, $\mathcal{J}(u, v)$ implies that $v < u^u$;

(b) for every k there are u and v such that $\mathcal{J}(u,v)$ and $v > u^k$.

Find a generalized Diophantine representation of exponentiation, given that there exists a Diophantine relation of exponential growth.

9. The conditions of Exercise 2.8 can be weakened even further. Let us say that a binary relation $\mathcal{R}$ *has roughly exponential growth* if there is a number m such that

(a) for every u and v, $\mathcal{R}(u,v)$ implies that

$$v < u^{u^{\cdot^{\cdot^{\cdot^{u}}}}},$$

where the tower of exponents is of height m.
(b) for every k there are u and v such that $\mathcal{R}(u,v)$ and $v > u^k$.

Show that exponentiation cannot be non-Diophantine if some relation of roughly exponential growth is Diophantine. (Note that here it is not required to find a corresponding generalized Diophantine representation.)

10. Show that if the equation

$$9(u^2 + 7v^2)^2 - 7(r^2 + 7s^2)^2 = 2$$

has only finitely many solutions, then there is a Diophantine relation of exponential growth.

Open questions

1. Is there a direct method for transforming a Diophantine relation of roughly exponential growth into a Diophantine relation of exponential growth?

2. Does the equation of Exercise 2.10 have an infinite number of solutions?

3. In a manner similar to (2.2.7), the kth-order recurrence relation

$$\delta(n+k) = b_{k-1}\delta(n+k-1) + \cdots + b_0\delta(n)$$

can be transformed into a first-order relation among matrices of order $k \times k$. If $b_0 = \pm 1$ then, as with (2.2.10), the determinants of the corresponding matrices are equal to $\pm c$ where the constant c is determined by $\delta(0), \ldots, \delta(k-1)$. As with (2.2.11), this condition can be stated in the form of a relation among the quantities $\delta(n), \ldots, \delta(n+k-1)$. When is it the case that this relation characterizes the sequence, so that (as in the case of (2.2.12)) it furnishes a Diophantine equation all solutions of which are among the terms of the sequence defined by the given recurrence?

Commentary

As was stated in the Commentary to Chapter 1, the origin of systematic investigations of the class of Diophantine sets was connected with Tarski's conjecture that the set of all powers of 2 is not Diophantine. When Julia Robinson did not succeed in proving this, she began to incline to the conjecture that exponentiation is Diophantine. In an important paper [1952], she gave sufficient conditions for exponentiation to be Diophantine. In particular, she showed that it would be sufficient to find a Diophantine relation of exponential growth (see Exercise 2.8) or at least roughly exponential growth (see Exercise 2.9). Relations of exponential growth are also known as *Julia Robinson predicates* (see, for example, Davis [1963]).

Later, Robinson [1969a] found various conditions sufficient for the existence of Diophantine relations of exponential growth; namely, she showed that it would be sufficient to find an infinite Diophantine set consisting entirely of primes, or to show that the set of all powers of 2 is Diophantine.

Davis [1968] found another sufficient condition consisting in the uniqueness of the trivial solution $u = r = 1, \quad v = s = 0$ of the equation from Exercise 2.10. However, Herrman [1971] established the existence of a non-trivial solution, and Shanks [1972], using a computer, also found a non-trivial solution:

$$u = 525692038369576, \qquad r = 2484616164142152,$$
$$v = 1556327039191013, \qquad s = 1381783865776981.$$

Nevertheless, as was mentioned in Davis, Matiyasevich, and Robinson [1976], this doesn't entirely spoil Davis's idea, because in fact it would suffice to show that the equation has only finitely many solutions (see Exercise 2.10).

These approaches have so far not led to success. Nevertheless, they remain of interest even after exponentiation was proved to be Diophantine in another way. This is because of the connection between these approaches and the so-called singlefold Diophantine representations (see Section 7.2 and the Commentary to Chapter 7). In addition to the conditions mentioned above, various other, more involved, conditions were proposed that also imply that exponentiation is Diophantine (see, for example, Davis [1962, 1966], Davis and Putnam [1958], Matiyasevich [1968b]).

The very first example of a Diophantine relation of exponential growth was published by Matiyasevich [1970]. It was the relation

$$v = \phi_{2u},$$

where ϕ_0, ϕ_1, ... are the Fibonacci numbers defined by

$$\phi_0 = 0, \qquad \phi_1 = 1, \qquad \phi_{n+2} = \phi_n + \phi_{n-1}.$$

According to the above-mentioned criterion due to Julia Robinson, this implied that exponentiation was Diophantine. Chronologically, this example of a Diophantine relation of exponential growth turned out to be the last missing link in establishing the algorithmic unsolvability of Hilbert's Tenth Problem, because the algorithmic unsolvability of exponential Diophantine equations had previously been established (for more details see the Commentary to Chapter 5).

The Fibonacci numbers are a special case (namely $b = 1$) of the sequences γ_b from Exercise 2.5, which are closely related to our α_b. All of these sequences have similar properties that can be used to prove that they are Diophantine. In the case of the sequences γ_b, this was shown by Chudnovsky [1970, 1971, 1984], Davis [1971, 1973a], Kosovskiĭ [1971], and also by Simon Kochen (see Davis [1971]) and Kurt Schütte (see Robinson [1971] or Fenstad [1971]).

In Section 2.3 we did a bit more than simply finding a Diophantine relation of exponential growth, as was done in Matiyasevich [1970]. The system (2.3.12)–(2.3.24) defines a relation among three numbers (rather than two), and if this relation holds, then the numbers satisfy two-sided inequalities stronger than those required in the definition of a relation of exponential growth. This opens a somewhat shorter path (used in Section 2.4) to proving that exponentiation is Diophantine than would be obtained from a straightforward application of the Julia Robinson criterion. Yet another method, also originating from Robinson [1952], is outlined in Exercise 2.7. Diophantine representations of b^c are presented in full detail in particular by Davis [1971], Kosovskii [1971], Matiyasevich [1971a, 1971b], and Matiyasevich and Robinson [1975].

Diophantine equations of the type (2.5.4) were explicitly presented by Ruohonen [1972] and Baxa [1993]. It is highly unlikely that transforming Fermat's simple exponential Diophantine equation to an equivalent complicated Diophantine equation could be of any help in investigations on Fermat's Last Theorem. On the other hand, this reduction can be viewed as an informal "psychological" argument in favor of the unsolvability of Hilbert's Tenth Problem, because otherwise the *process* required by the problem would permit one, in particular, to determine whether the Theorem is true or false.

Chudnovsky [1971] (cf. [1984]) states that Davis's question about the solvability of the equation from Exercise 2.10 "can be reduced to studying the arithmetical properties of the sequences (A_n, B_n) of solutions of the equation $9x^2 - 7y^2 = 2$" and by "studying these sequences one can obtain a Diophantine representation for $y = 2^x$ with $x > C$ where C is a constant. The result obtained answers Davis's question." Today we know that Davis's conjecture that the trivial solution is unique is not true, and it is not clear what Chudnovsky had in mind: did he mean that the number of solutions was finite or did he propose to use the sequences (A_n, B_n) in some other way? That is why Question 2.2 is stated as being open.

An answer to Open Question 2.3 will most likely be connected with an analysis of the multiplicative group of units of the field $\mathbb{Q}(\chi)$, where

$$\chi^k = b_{k-1}\chi^{k-1} + \cdots + b_0,$$

and most likely for the answer to be positive it is necessary that $b_0 = \pm 1$ and $k \leq 4$, and also in the case $k = 4$ that the equation have no real roots, while in the case $k = 3$ that it have only one real root (because, by Dirichlet's Theorem, it is only under these conditions that the field has a unique fundamental unit).

3 Diophantine Coding

Any particular Diophantine equation has a fixed number of unknowns. Nevertheless, Diophantine equations can be used to deal with finite sequences of numbers (or *tuples*, as we shall refer to them) of arbitrary length. In this chapter we develop the appropriate techniques.

3.1 Cantor numbering

Cantor's classic proof that the countable union of countable sets is countable is based on the following linear ordering of all pairs of natural numbers:

$$\langle 0,0\rangle, \langle 0,1\rangle, \langle 1,0\rangle, \langle 0,2\rangle, \langle 1,1\rangle, \langle 2,0\rangle, \langle 0,3\rangle, \dots, \langle 3,0\rangle \dots \tag{3.1.1}$$

We shall frequently take advantage of the fortunate fact that the number of the pair $\langle a, b\rangle$ in the sequence (3.1.1) is given by a polynomial in a and b, namely,

$$\mathrm{Cantor}(a,b) = \frac{(a+b)^2 + 3a + b}{2}. \tag{3.1.2}$$

(We begin the numbering of pairs in (3.1.1) with 0 rather than 1.) Thus, we have the Diophantine functions $\mathrm{Elema}(c)$ and $\mathrm{Elemb}(c)$ that, given the number of a pair, yield its first and second elements, respectively:

$$a = \mathrm{Elema}(c) \iff \exists y\,[(a+y)^2 + 3a + y = 2c], \tag{3.1.3}$$

$$b = \mathrm{Elemb}(c) \iff \exists x\,[(x+b)^2 + 3x + b = 2c]. \tag{3.1.4}$$

The numbering of pairs can easily be generalized to a numbering of triples, quadruples, and so on. For example, one can define

$$\begin{aligned} \mathrm{Cantor}_1(a_1) &= a_1 \\ \mathrm{Cantor}_{n+1}(a_1,\dots,a_{n+1}) &= \mathrm{Cantor}_n(a_1,\dots,a_{n-1},\mathrm{Cantor}(a_n,a_{n+1})) \end{aligned} \tag{3.1.5}$$

and call $\mathrm{Cantor}_n(a_1,\dots,a_n)$ the *Cantor number* of tuple $\langle a_1,\dots,a_n\rangle$. Analogously to (3.1.3) and (3.1.4), we have the Diophantine function $\mathrm{Elem}_{n,m}(c)$, which yields the mth element of the n-tuple with Cantor number c:

$$a = \mathrm{Elem}_{n,m}(c) \iff \tag{3.1.6}$$

$$\exists x_1 \dots x_{m-1} x_{m+1} \dots x_n\,[2^{2^n}\,\mathrm{Cantor}_n(x_1,\dots,x_{m-1},a,x_{m+1},\dots,x_n) = 2^{2^n} c].$$

(The factor 2^{2^n} in (3.1.7) is introduced because the coefficients of Cantor_n are not integers.)

3.2 Gödel coding

The numbering of tuples of natural numbers introduced in the previous section has a crucial disadvantage for us. Although the inverse function $\mathrm{Elem}_{n,m}(c)$ is a Diophantine function of one argument for any fixed values of n and m, there is no easy way to prove that $\mathrm{Elem}_{n,m}(c)$ is a Diophantine function of the three arguments c, m, and n. In order to work with tuples whose lengths are not fixed in advance, we need to use other methods.

One such method is based on the Chinese Remainder Theorem (see the Appendix). Consider an arbitrary tuple

$$\langle a_1, \ldots, a_n \rangle. \tag{3.2.1}$$

Let

$$b_1, \ldots, b_n \tag{3.2.2}$$

be any pairwise coprime numbers such that

$$a_i < b_i, \qquad i = 1, \ldots, n. \tag{3.2.3}$$

By the Chinese Remainder Theorem, we can find a number a such that

$$a_i = \mathrm{rem}(a, b_i), \quad i = 1, \ldots, n. \tag{3.2.4}$$

Thus all the elements of tuple (3.2.1) are uniquely determined by the numbers

$$a, \quad b_1, \ldots, b_n. \tag{3.2.5}$$

At first sight we haven't gained anything by proceeding from the n numbers (3.2.1) to the $n+1$ numbers (3.2.5). However, the benefit is as follows: in contrast to (3.2.1), we have great freedom in choosing the numbers (3.2.2), and we can choose them to be determined by just a few numbers. For example, put

$$b_i = bi + 1, \quad i = 1, \ldots, n, \tag{3.2.6}$$

where b is a multiple of $n!$ and is large enough to imply the inequalities (3.2.3). Then the numbers b_i and b_j are coprime when $i < j$. In fact, suppose that

$$p \mid bi + 1, \tag{3.2.7}$$

$$p \mid bj + 1. \tag{3.2.8}$$

Then clearly

$$\gcd(p, b) = 1, \tag{3.2.9}$$
$$\gcd(p, n!) = 1; \tag{3.2.10}$$

however,

$$p \mid b(j - i), \tag{3.2.11}$$

and hence

$$p \mid j - i, \tag{3.2.12}$$

so that

$$p \leq j - i < n. \tag{3.2.13}$$

It follows from (3.2.10) and (3.2.13) that $p = 1$; i.e., b_i and b_j are coprime.

Thus, the pair $\langle a, b \rangle$ contains complete information about the individual elements of tuple (3.2.1); however, it contains no hint as to its length. We can include this additional information quite formally; namely, we call the triple $\langle a, b, c \rangle$ a *Gödel code* of tuple (3.2.1) if $c = n$, and for $i = 1, \ldots, n$,

$$a_i = \mathrm{GElem}(a, b, i), \tag{3.2.14}$$

where GElem is the Diophantine function such that

$$e = \mathrm{GElem}(a, b, i) \iff e = \mathrm{rem}(a, bi + 1). \tag{3.2.15}$$

Naturally, $\langle a, b, 0 \rangle$ is the code of the *empty tuple*, i.e., the tuple of length 0, which contains no elements. Of course, instead of the triple $\langle a, b, c \rangle$, we could employ the single number $\mathrm{Cantor}_3(a, b, c)$ as our Gödel code, thus obtaining a numbering of tuples of arbitrary length by natural numbers.

We should emphasize one important difference between the Gödel and the Cantor codings. While the Cantor number of a tuple is unique, each tuple has infinitely many Gödel codes, due to the arbitrariness in the choice of b and the consequent arbitrariness in the choice of a. This is not important: even if we fix one particular code for each tuple, we cannot have a Diophantine function that would yield the code of $\langle a_1, \ldots, a_n \rangle$ from $a_1, \ldots, a_n$, because a Diophantine function has only a fixed number of arguments.

3.3 Positional coding

Gödel coding permits us to work with tuples of indefinite length, and any particular element of a tuple is easily determined from the code and the position of the element in the tuple by a simple Diophantine function. However, it is much more difficult to show that other relations and functions are Diophantine, for example, *concatenation*, which from codes of the tuples

$$\langle p_1, \ldots, p_m \rangle \tag{3.3.1}$$

and

$$\langle q_1, \ldots, q_n \rangle \tag{3.3.2}$$

yields a code of the tuple

$$\langle p_1, \ldots, p_m, q_1, \ldots, q_n \rangle. \tag{3.3.3}$$

For the coding to be introduced in this section, we shall be able to establish that concatenation and various other functions and relations are Diophantine.

The new code consists of three numbers a, b, c, where c has the same value as in Gödel coding; i.e., it is equal to n, the length of the coded tuple

$$\langle a_1, \ldots, a_n \rangle. \tag{3.3.4}$$

The second number, b, is now required to satisfy the stronger inequality

$$b > a_i, \quad i = 1, \ldots, n, \tag{3.3.5}$$

and the value of a is defined by

$$a = a_n b^{n-1} + a_{n-1} b^{n-2} + \cdots + a_1 b^0. \tag{3.3.6}$$

In other words, a_n, $\ldots$, a_1 are nothing other than the digits in the b-ary representation of a, and hence tuple (3.3.4) is uniquely determined by the numbers a, b, and c. A triple $\langle a, b, c \rangle$ is called a *positional code* of tuple (3.3.4) if $c = n$ and (3.3.5) and (3.3.6) hold. Clearly, triple $\langle 0, b, 0 \rangle$ is a code of the empty tuple.

As with Gödel coding, every tuple has infinitely many codes, but in contrast to Gödel coding, not every triple is a positional code of a tuple. This is not a great disadvantage because we can easily establish that the relation *is a positional code* is Diophantine:

$$\text{Code}(a, b, c) \iff b \geq 2 \,\&\, a < b^c. \tag{3.3.7}$$

In what follows, we shall deal mainly with positional codes, and the word "positional" will be often omitted. For the elements of a code $\langle a, b, c\rangle$, the following terms will be used: a is the *cipher*, b is the *base*, c is the *length*.

It is easy to see that the function $\mathrm{Elem}(a, b, d)$ yielding the dth element of the corresponding tuple is Diophantine:

$$e = \mathrm{Elem}(a, b, d)$$
$$\iff \exists xyz\,[d = z + 1 \,\&\, a = xb^d + eb^z + y \,\&\, e < b \,\&\, y < b^z]. \quad (3.3.8)$$

Let $\mathrm{Concat}(a, b, c, a_1, b_1, c_1, a_2, b_2, c_2)$ be the relation: *triple* $\langle a, b, c\rangle$ *is a code of the concatenation of the tuples with codes* $\langle a_1, b_1, c_1\rangle$ *and* $\langle a_2, b_2, c_2\rangle$. Then it is easy to see that

$$\mathrm{Concat}(a, b, c, a_1, b, c_1, a_2, b, c_2)$$
$$\iff \mathrm{Code}(a_1, b, c_1) \,\&\, \mathrm{Code}(a_2, b, c_2) \,\&\, a = a_2 b^{c_1} + a_1 \,\&\, c = c_1 + c_2. \quad (3.3.9)$$

Thus, when the base is fixed, concatenation is a Diophantine operation. Actually, since we are considering only functions whose values are single natural numbers, it would be more correct to say that there is a pair of Diophantine functions that respectively yield the cipher and the length of the concatenation of two tuples. However, rather than introducing notation for these functions, we shall simply write $\langle a_1, b, c_1\rangle + \langle a_2, b, c_2\rangle$ for the triple that codes (with base b) the tuple resulting from concatenating the tuples with codes $\langle a_1, b, c_1\rangle$ and $\langle a_2, b, c_2\rangle$.

In Section 3.5 we shall establish that Concat is Diophantine in a general setting, i.e., when all three bases may be different.

3.4 Binomial coefficients, the factorial, and the prime numbers are Diophantine

As a somewhat unexpected corollary to the simple coding technique developed in the previous section, we can see that binomial coefficients are Diophantine. For this it suffices to note that the triple $\langle (b+1)^n, b, n+1\rangle$ is a code of the tuple

$$\left\langle \binom{n}{0}, \binom{n}{1}, \ldots, \binom{n}{n} \right\rangle \quad (3.4.1)$$

provided, of course, that b is sufficiently large, for example, when $b = 2^n + 1$. Hence,

$$c = \binom{n}{m} \iff c = \mathrm{Elem}((2^n+2)^n, 2^n+1, m+1). \quad (3.4.2)$$

For binomial coefficients we have the well-known expression in terms of the factorial:

$$\binom{n}{m} = \frac{n!}{m!\,(n-m)!}. \tag{3.4.3}$$

We can invert this formula and express the factorial in terms of binomial coefficients and exponentiation:

$$\begin{aligned} m! &= \frac{n!}{\binom{n}{m}(n-m)!} \\ &= \frac{n(n-1)\dots(n-m+1)}{\binom{n}{m}} \\ &= \frac{n^m}{\binom{n}{m}}\left(1-\frac{1}{n}\right)\dots\left(1-\frac{m-1}{n}\right) \\ &= \lim_{n\to\infty} \frac{n^m}{\binom{n}{m}}\left(1-\frac{1}{n}\right)\dots\left(1-\frac{m-1}{n}\right) \\ &= \lim_{n\to\infty} \frac{n^m}{\binom{n}{m}}. \end{aligned} \tag{3.4.4}$$

The operation lim doesn't belong to our Diophantine language, but as in Section 2.4, we can use the function div as a substitute. Namely, (3.4.4) implies that for n sufficiently large,

$$m! = n^m \operatorname{div} \binom{n}{m}. \tag{3.4.5}$$

It is not difficult to check that (3.4.5) is valid when $n \geq (m+1)^{m+2}$, so that

$$m! = (m+1)^{m(m+2)} \operatorname{div} \binom{(m+1)^{m+2}}{m}. \tag{3.4.6}$$

Once we have proved that the factorial is Diophantine, we can readily prove that the property *is a prime number* is Diophantine:

$$\operatorname{Prime}(a) \iff a > 1 \,\&\, \gcd(a, (a-1)!) = 1. \tag{3.4.7}$$

As was shown in Section 1.4, this implies that *there is a polynomial in several variables with integer coefficients with the following property: the set of all natural number values that the polynomial assumes when its variables are permitted to range over the natural numbers is precisely the set of all prime numbers.*

3.5 Comparison of tuples

As was mentioned above, every tuple has infinitely many codes, corresponding to the various bases; two codes for the same tuple will be referred to as *equivalent.* The main purpose of this section is to show that this relation of equivalence, i.e., the relation *triples* $\langle a_1, b_1, c_1\rangle$ *and* $\langle a_2, b_2, c_2\rangle$ *are codes of the same tuple*, is Diophantine; this relation will be denoted by $\text{Equal}(a_1, b_1, c_1, a_2, b_2, c_2)$. In addition, we show that various other relations of comparison are Diophantine. In particular, the relation of element-wise inequality

$$\text{NotGreater}(a_1, b_1, a_2, b_2) \iff \forall k\,[\text{Elem}(a_1, b_1, k) \leq \text{Elem}(a_2, b_2, k)] \quad (3.5.1)$$

and the relation

$$\text{Small}(a, b, c, e) \iff \text{Code}(a, b, c)\ \&\ \forall k\,[\text{Elem}(a, b, k) \leq e], \quad (3.5.2)$$

which compares elements of a tuple with a particular number, are Diophantine.

Also, we shall establish that a certain auxiliary relation Eq is Diophantine. This relation will be stronger than Equal, in the sense that

$$\text{Eq}(a_1, b_1, c_1, a_2, b_2, c_2) \Longrightarrow \text{Equal}(a_1, b_1, c_1, a_2, b_2, c_2). \quad (3.5.3)$$

We will be able to claim the converse implication only when b_2 is a prime number sufficiently large with respect to b_1 and c_1. Nevertheless, this will be enough to enable us to express Equal in terms of Eq as follows:

$$\text{Equal}(a_1, b_1, c_1, a_2, b_2, c_2) \iff$$
$$\exists xyz\,[\text{Eq}(a_1, b_1, c_1, x, y, z)\ \&\ \text{Eq}(a_2, b_2, c_2, x, y, z)]. \quad (3.5.4)$$

We begin by proving that another auxiliary relation is Diophantine, namely, the relation

$$\text{PNotGreater}(a_1, a_2, b) \iff \text{Prime}(b)\ \&\ \text{NotGreater}(a_1, b, a_2, b). \quad (3.5.5)$$

The use of *one single prime* base for both tuples enables us to apply Kummer's Theorem (see the Appendix), thus obtaining the following generalized Diophantine representation:

$$\text{PNotGreater}(a_1, a_2, b) \iff \text{Prime}(b)\ \&\ b \nmid \binom{a_2}{a_1}. \quad (3.5.6)$$

Using PNotGreater, we can show that the relation

$$\text{PSmall}(a, b, c, e) \iff \text{Prime}(b)\ \&\ \text{Small}(a, b, c, e) \quad (3.5.7)$$

is also Diophantine. Clearly,

$$\text{PSmall}(a,b,c,e) \iff \text{Prime}(b)\ \&\ [e \geq b \vee \text{PNotGreater}(a, \text{Repeat}(e,b,c), b)], \tag{3.5.8}$$

where $\text{Repeat}(p,q,r) = p(q^r - 1)/(q-1)$ is the cipher of the tuple $\langle p, \ldots, p\rangle$ of length r to the base q (provided, of course, that $p < q$).

Now we have enough tools to tackle our main goal. Suppose we have two bases, "small" b_1 and "large" b_2, and the code $\langle a_2, b_2, c\rangle$ of some tuple, all of whose elements are less than b_1. Clearly, the cipher a_1 of this same tuple to the base b_1 satisfies the congruence

$$a_1 \equiv a_2 \pmod{b_2 - b_1} \tag{3.5.9}$$

and the inequality

$$a_1 < b_1^c, \tag{3.5.10}$$

and hence if b_2 is so large that

$$b_1^c < b_2 - b_1, \tag{3.5.11}$$

then a_1 is uniquely determined by conditions (3.5.9) and (3.5.10). Now we are in a position to define the relation Eq:

$$\begin{aligned} \text{Eq}(a_1,b_1,c_1,a_2,b_2,c_2) \iff & \text{Code}(a_1,b_1,c_1)\ \&\ c_1 = c_2 \\ & \&\ \text{PSmall}(a_2,b_2,c_2,b_1-1)\ \&\ b_1^{c_1} + b_1 < b_2\ \&\ a_1 \equiv a_2 \pmod{b_2 - b_1}. \end{aligned} \tag{3.5.12}$$

It is easy to see that with this definition of Eq, implication (3.5.3) and equivalence (3.5.4) are valid, which proves that Equal is a Diophantine relation.

Now that we have at our disposal the relations Equal, PNotGreater, and PSmall, we can easily show that the relations NotGreater and Small are also Diophantine by passing to equivalent codes with prime bases:

$$\begin{aligned} \text{NotGreater}(a_1,b_1,a_2,b_2) \iff \exists x_1 x_2 y z\, [& \text{Equal}(a_1,b_1,z,x_1,y,z) \\ & \&\ \text{Equal}(a_2,b_2,z,x_2,y,z) \\ & \&\ \text{PNotGreater}(x_1,x_2,y)], \end{aligned} \tag{3.5.13}$$

$$\text{Small}(a,b,c,e) \iff \exists xy\, [\text{Equal}(a,b,c,x,y,c)\ \&\ \text{PSmall}(x,y,c,e)]. \tag{3.5.14}$$

In a similar way we can establish that the nine place relation Concat introduced in Section 3.3 is Diophantine:

$$\begin{aligned}\text{Concat}(a, b, c, a_1, b_1, c_1, a_2, b_2, c_2) \iff \exists x_1 x_2 \, [&\text{Equal}(a_1, b_1, c_1, x_1, b, c_1) \\ &\& \ \text{Equal}(a_2, b_2, c_2, x_2, b, c_2) \\ &\& \ a = x_2 b^{c_1} + x_1 \\ &\& \ c = c_1 + c_2]\end{aligned} \tag{3.5.15}$$

(cf. (3.3.9)).

3.6 Extensions of functions to tuples

In this section we are going to work with positional codes with some fixed base $b \geq 3$. The elements of the corresponding tuples thus belong to the set

$$\mathfrak{N}_b = \{0, 1, \ldots, b-1\}.$$

Let F be a function defined on $\mathfrak{N}_b$ with values also lying in $\mathfrak{N}_b$. Such a function generates a new function, denoted by $F[b]$, which is defined as follows: if $\langle a, b, c \rangle$ is the code of the tuple

$$\langle a_1, \ldots, a_c \rangle, \tag{3.6.1}$$

then the value of $F[b](a, c)$ is equal to the cipher of tuple

$$\langle F(a_1), \ldots, F(a_c) \rangle. \tag{3.6.2}$$

(If a is not the cipher of any tuple of length c to the base b, then $F[b](a, c)$ is left undefined.) We are going to show that *the function $F[b]$ is Diophantine.*

For the tuple (3.6.1) we can construct b *characteristic tuples*

$$\begin{gathered}\langle h_{0,1}, \ldots, h_{0,c} \rangle, \\ \vdots \\ \langle h_{b-1,1}, \ldots, h_{b-1,c} \rangle,\end{gathered} \tag{3.6.3}$$

where

$$h_{k,j} = \begin{cases} 1 & \text{if } a_j = k, \\ 0 & \text{otherwise.} \end{cases} \tag{3.6.4}$$

Let $h_0, \ldots, h_{b-1}$ be the ciphers of the tuples (3.6.3). Clearly,

$$0 \cdot h_0 + 1 \cdot h_1 + \cdots + (b-1) \cdot h_{b-1} = a. \tag{3.6.5}$$

Also,

$$h_0 + \cdots + h_{b-1} = \text{Repeat}(1, b, c) \tag{3.6.6}$$

and

$$\text{Ortnorm}_b(h_k, h_l, c), \quad 0 \leq k < l < b, \tag{3.6.7}$$

where $\text{Ortnorm}_b(q_1, q_2, r)$ is the orthonormality relation: *all elements of the tuples with codes* $\langle q_1, b, r\rangle$ *and* $\langle q_2, b, r\rangle$ *are no greater than* 1 *and* 1 *does not occur in the same position in both tuples.* This relation is Diophantine because

$$\begin{aligned}&\text{Ortnorm}_b(q_1, q_2, r)\\ &\qquad \Longleftrightarrow \text{Small}(q_1, b, r, 1) \,\&\, \text{Small}(q_2, b, r, 1) \,\&\, \text{Small}(q_1 + q_2, b, r, 1).\end{aligned} \tag{3.6.8}$$

It is easy to see that conditions (3.6.5)–(3.6.7) uniquely determine the ciphers of the characteristic tuples for given a, b, and c. These conditions are Diophantine, and to complete the proof that $F[b]$ is Diophantine, it suffices to note that the cipher of tuple (3.6.2) is equal to

$$F(0) \cdot h_0 + \cdots + F(b-1) \cdot h_{b-1}. \tag{3.6.9}$$

Note that this technique can easily be generalized to the case where F is a function of m arguments. Then, the arguments of $F[b]$ would be the length c and the ciphers $a_1, \ldots, a_m$ of given tuples

$$\begin{gathered}\langle a_{1,1}, \ldots, a_{1,c}\rangle,\\ \vdots\\ \langle a_{m,1}, \ldots, a_{m,c}\rangle.\end{gathered} \tag{3.6.10}$$

To obtain an analog of (3.6.9), we need b^m characteristic tuples defined analogously to (3.6.4) by

$$h_{k_1,\ldots k_m,l} = \begin{cases} 1 & \text{if } a_{1,l} = k_1, \ldots, a_{m,l} = k_m, \\ 0 & \text{otherwise.} \end{cases} \tag{3.6.11}$$

The ciphers of the tuples $h_{k_1,\ldots,k_m,l}$ satisfy conditions analogous to conditions (3.6.5)–(3.6.7) and are uniquely determined by them.

Exercises

1. In most instances when the polynomials Cantor_n are used in this book, it is not necessary that they assume all natural numbers as values. In fact, instead of Cantor_n, one could use any polynomial that assumes each of its values only once. Show that the polynomial

$$a_1 + (a_1 + a_2)^2 + (a_1 + a_2 + a_3)^3 + \cdots + (a_1 + \cdots + a_n)^n$$

has this property.

2. Show that $\text{Cantor}(a, b)$ is the unique polynomial of degree less than 5 with rational coefficients (to within the order of its arguments) that maps the set of all pairs of natural numbers one-to-one to the natural numbers.

3. Show that $\prod_{k \leq c}(a + bk)$ is a Diophantine function of three arguments.

4. Find a unary exponential Diophantine representation for the central binomial coefficient $\binom{2n}{n}$ using the identity

$$\frac{1}{\sqrt{(1-4\chi)}} = \sum_{n=0}^{\infty} \binom{2n}{n} \chi^n,$$

which is valid for $|\chi| < \frac{1}{4}$.

5. Find an exponential Diophantine representation for the factorial using the classical identities

$$\sum_{k=0}^{\infty} \frac{\tau^k}{k!} = e^{\tau} = \lim_{y \to \infty} \left(1 + \frac{\tau}{y}\right)^y.$$

6. Show that $a/b = |B_{2c}|$ is a Diophantine relation among a, b, and c, where $B_0 = 1$, $B_1 = -1/2$, $B_2 = 1/6$, ... are the Bernoulli numbers.

7. Find a generalized Diophantine representation for the property Prime under the assumption that the binary relation

$$\text{Bernoulli}(a, b) \iff \exists n \left[\frac{a}{b} = |B_n|\right]$$

is Diophantine.

8. Show that the negations of the relations Code, Equal, NotGreater, and Small are Diophantine.

9. Let Mult and Add be the functions that, given codes $\langle p, q, r\rangle$ and $\langle u, v, w\rangle$ of tuples $\langle p_1, \ldots, p_r\rangle$ and $\langle u_1, \ldots, u_w\rangle$, return the ciphers to the base qv of the tuples

$$\langle p_1u_1, \ldots, p_1u_w, \ldots, p_ru_1, \ldots, p_ru_w\rangle$$

and

$$\langle p_1 + u_1, \ldots, p_1 + u_w, \ldots, p_r + u_1, \ldots, p_r + u_w\rangle,$$

respectively. Show that Mult and Add are Diophantine.

10. Let Sum be the function that, given a code of a tuple, returns the sum of the elements of the tuple. Show that Sum is Diophantine.

11. Let $\mathrm{Less}(a_1, a_2, b)$ be the cipher of the tuple $\langle t_1, \ldots, t_c\rangle$ to the base b, where $t_i = 1$ if $\mathrm{Elem}(a_1, b, i) < \mathrm{Elem}(a_2, b, i)$ and $t_i = 0$ otherwise. Show that Less is Diophantine.

12. Show that π is a Diophantine function where, as usual, $\pi(a)$ denotes the number of primes no greater than a.

13. Construct a polynomial $\mathrm{Primenth}(n, x_1, \ldots, x_m)$ that, for each fixed n and arbitrary values of $x_1, \ldots, x_m$, assumes a unique positive value, namely, the nth prime number.

14. Our proof in Section 3.5 that Equal, NotGreater, and Small are Diophantine took advantage of the fact (proved in Section 3.4) that Prime was already known to be Diophantine. Show that we could avoid using this fact by employing only the $p = 2$ case of Kummer's Theorem.

15. Let

$$\delta(0) = a_0, \ldots, \delta(k-1) = a_{k-1}, \qquad \delta(n+k) = b_{k-1}\delta(n+k-1) + \cdots + b_0\delta(n)$$

be a kth-order linear recurrent sequence. Show that δ is a Diophantine function of the $2k+1$ arguments $a_0, \ldots, a_{k-1}, b_0, \ldots, b_{k-1}, n$.

16. The technique of Section 3.6 will find its main application in Chapter 5, where we shall work with codes to a fixed base. However, some functions can easily be

managed without fixing the base. Let Min be the ternary function such that

$$\forall k\big[\mathrm{Elem}\big(\mathrm{Min}(a_1,a_2,b),b,k\big)=\min\big(\mathrm{Elem}(a_1,b,k),\mathrm{Elem}(a_2,b,k)\big)\big].$$

Show that Min is Diophantine.

17. Show that if $ad-bc=1$, then the matrix $\left(\begin{smallmatrix}a&b\\c&d\end{smallmatrix}\right)$ can be represented as a product of the matrices $\left(\begin{smallmatrix}1&1\\0&1\end{smallmatrix}\right)$ and $\left(\begin{smallmatrix}1&0\\1&1\end{smallmatrix}\right)$ and that this representation is unique. This fact can be used to code tuples of 0's and 1's of arbitrary length. Show that under the coding defined in this manner, concatenation is Diophantine.

Open questions

1. Is there an analog to Kummer's Theorem (see the Appendix) for the case of the positional system in which the weights of the digits are equal to $\alpha_b(1)$, $\alpha_b(2)$, ...?

2. By Wolstenholme's Theorem [1862], for $n>3$

$$\mathrm{Prime}(n)\Longrightarrow\binom{2n-1}{n-1}\equiv 1\pmod{n^3}.$$

Is the converse implication true?

Commentary

The use of the Chinese Remainder Theorem for coding tuples of arbitrary length was first introduced by Kurt Gödel in his famous paper [1931] in which he showed that for any "reasonable" formal system there are arithmetic statements that can neither be proved nor disproved in that system. In this same paper, he also introduced a coding based on the *Fundamental Theorem of Arithmetic*, i.e., the uniqueness of prime factorization. Gödel used these codings for arithmetization, and so he was free to use arbitrary quantifiers. In Diophantine representations one can use only existential quantifiers, and that is why we are more restricted in the choice of available codings.

Gödel coding has played a special role in research on Hilbert's Tenth Problem, and not only because of the simplicity of the function GElem. As opposed to other codings, the Chinese Remainder Theorem furnishes us with a unique opportunity to calculate the values of a polynomial, as follows: Let $P(x_1,\dots,x_m)$ be a polynomial whose coefficients are natural numbers, and let $\langle a_1,b,c\rangle$, ..., $\langle a_m,b,c\rangle$ be Gödel codes of the tuples

$$\langle a_{1,1}, \dots, a_{1,c} \rangle, \dots, \langle a_{m,1}, \dots, a_{m,c} \rangle.$$

Then the triple $\langle P(a_1, \dots, a_m), b, c \rangle$ is a Gödel code of the tuple

$$\langle P(a_{1,1}, \dots, a_{m,1}), \dots, P(a_{1,c}, \dots, a_{m,c}) \rangle,$$

provided that b is sufficiently large. Taking advantage of this property, Davis, Putnam, and Robinson [1961] achieved essential progress towards the unsolvability of Hilbert's Tenth Problem. Namely, they established the algorithmic unsolvability of exponential Diophantine equations (for more details, see the Commentary to Chapters 5 and 6).

Gödel coding has some disadvantages, too. While the function GElem has a very simple Diophantine representation, it is much more difficult to show that other operations on codes—concatenation, for example—are Diophantine. In this respect, positional coding "redistributes" the work: to show that Elem is Diophantine, we needed to first establish that exponentiation is Diophantine. However, once this was accomplished, it was relatively easy to show that many other functions and relations are Diophantine as well. Before it had been proved that exponentiation is Diophantine, another type of positional coding was used in research on Hilbert's Tenth Problem, namely, one with the weights of the digits forming a second-order recurrent sequence (see Matiyasevich [1968a, 1968b]), and Open Question 3.1 is caused by such a coding.

Soon after proving that exponentiation is Diophantine, Matiyasevich [1971b] used positional coding to construct Diophantine representations of higher-order linear recurrent sequences (see Exercise 3.15). When the possibilities inherent in Kummer's Theorem came to be understood, positional coding was revealed to be an even more powerful tool for constructing Diophantine representations. This new technique was employed for the first time in Matiyasevich [1976].

The fact that binomial coefficients and the factorial are exponential Diophantine was proved by Julia Robinson [1952] without any reference to the general concept of positional coding.

In the same paper, Robinson gave an exponential Diophantine representation for the set of all prime numbers. Later, Putnam [1960] noted that a Diophantine set is the positive part of the range of a polynomial (see Section 1.4). Thus, it became clear that if exponentiation were established to be Diophantine, it would become possible to construct a polynomial such that the positive values it assumed would coincide precisely with the prime numbers. At the time, this corollary seemed rather unlikely, and consequently many researchers regarded it as heuristic evidence that exponentiation was not in fact Diophantine.

Due to the importance of the set of primes, a number of papers were dedicated particularly to Diophantine representations of this set. In particular, many researchers were interested in knowing how few variables would suffice in a polynomial representing all primes and only primes. The first upper bound of 24 variables was obtained by Matiyasevich [1971a] (the bound was reduced to 21 variables in the appendix to the English translation). Later, the bound was further reduced to 12 variables by Wada [1975] and by Jones, Sato, Wada, and Wiens [1976]. At present (1992), the record is 10 variables, achieved by Matiyasevich [1977a]. The construction of this polynomial was based on specific properties of the primes. Another polynomial in 10 variables for the primes can be constructed based on a quite general technique. (See Matiyasevich [1977c], Jones [1982b], and the Commentary to Chapter 8.)

Of course, polynomials based on a general technique tend to be rather cumbersome. On the other hand, by using specific properties of primes, one can construct a polynomial occupying only a few lines. Jones, Sato, Wada and Wiens [1976] exhibited the following polynomial:

$$
\begin{aligned}
(k+2)\Big\{1 &- [wz+h+j-q]^2 \\
&- [(gk+2g+k+1)(h+j)+h-z]^2 \\
&- [2n+p+q+z-e]^2 \\
&- \left[16(k+1)^3(k+2)(n+1)^2+1-f^2\right]^2 \\
&- \left[e^3(e+2)(a+1)^2+1-o^2\right]^2 \\
&- \left[(a^2-1)y^2+1-x^2\right]^2 \\
&- \left[16r^2y^4(a^2-1)+1-u^2\right]^2 \\
&- [n+l+v-y]^2 \\
&- \left[\left((a+u^2(u^2-a))^2-1\right)(n+4dy)^2+1-(x+cu)^2\right]^2 \\
&- \left[(a^2-1)l^2+1-m^2\right]^2 \\
&- \left[q+y(a-p-1)+s(2ap+2a-p^2-2p-2)-x\right]^2 \\
&- \left[z+pl(a-p)+t(2ap-p^2-1)-pm\right]^2 \\
&- [ai+k+1-l-i]^2 \\
&- \left[p+l(a-n-1)+b(2an+2a-n^2-2n-2)-m\right]^2\Big\}.
\end{aligned}
$$

This polynomial contains 26 variables (all the letters of the alphabet!), and the set of its *positive* values is exactly the set of all prime numbers. Considering only positive values, one can use, instead of (1.4.8), a representation of the form

$$x_0(1 - D^2(x_0, \dots, x_m)) = a.$$

This minor modification leads to the following paradox: the polynomial written above, representing only primes, is itself the product of two polynomials!

Now that we have Diophantine representations for the set of all primes, we can easily find Diophantine representations for the set of all *Mersenne primes*, i.e., primes of the form $2^n - 1$, and for the set of all *Fermat primes*, i.e., odd primes of the form $2^n + 1$. However, for numbers of the form $2^n \pm 1$, there are special criteria for primality stated in terms of divisibility of elements of suitable recurrent sequences, which enabled Jones [1979] to provide Diophantine representations for these sets with only 7 variables. (At present, we know only five odd primes of the form 2^n+1, and it considered very likely that we know all of them; if this turns out to be the case, then the Fermat primes will be definable by a polynomial with a single variable.)

It is well known that the Mersenne primes are closely connected with even perfect numbers. Kryauchyukas [1979] obtained a Diophantine representation for the set of *all* perfect numbers.

A positive answer to Open Question 3.2 would enable us to provide a rather simple Diophantine representation for the set of all primes (cf. Exercise 3.4).

For a recent paper about the Wolstenholme Theorem, see Bauer [1988].

4 Universal Diophantine Equations

In this chapter we are going to construct a universal Diophantine equation. Solving any given Diophantine equation can be reduced to solving the universal equation by choosing suitable values for its parameters. The universal equation will enable us to construct a Diophantine set with a non-Diophantine complement.

4.1 Basic definitions

A universal equation has the form of a family of equations

$$U(a_1, \dots, a_n, k_1, \dots, k_l, y_1, \dots, y_w) = 0 \tag{4.1.1}$$

whose parameters are separated into the *element parameters* $a_1, \dots, a_n$ and the *code parameters* $k_1, \dots, k_l$. Equation (4.1.1) is called *universal* if for any given Diophantine equation with n parameters

$$D(a_1, \dots, a_n, x_1, \dots, x_m) = 0, \tag{4.1.2}$$

there are numbers $k_1^D, \dots, k_l^D$ such that equation (4.1.2) has a solution in $x_1, \dots, x_m$ for precisely those values of the parameters $a_1, \dots, a_n$ for which the equation

$$U(a_1, \dots, a_n, k_1^D, \dots, k_l^D, y_1, \dots, y_w) = 0 \tag{4.1.3}$$

has a solution in $y_1, \dots, y_w$.

In other words, we can say that equation (4.1.3) provides another representation of the Diophantine set defined by equation (4.1.2). Thus, every universal equation gives rise to a coding of the Diophantine sets of a certain fixed dimension; that is, the tuple $\langle k_1^D, \dots, k_l^D \rangle$ may be regarded as a *code of the set* defined by equation (4.1.2).

We can also treat equation (4.1.1) as an equation in which all the parameters $a_1, \dots, a_n, k_1, \dots, k_l$ are element parameters; then equation (4.1.1) defines a Diophantine set of $(l+n)$-tuples of natural numbers called, naturally enough, a *universal Diophantine set*.

It is easy to see that, at the cost of an increase in the number of unknowns, any universal equation can be transformed into another universal equation with the same number of element parameters but with a single code parameter. In fact, if (4.1.1) is a universal equation, then so is the equation

$$U^2(a_1, \dots, a_n, k_1, \dots, k_l, y_1, \dots, y_w) + (k - 2^{2^l}\,\mathrm{Cantor}_l(k_1, \dots, k_l))^2 = 0 \tag{4.1.4}$$

with one code parameter k and $l + w$ unknowns $k_1, \dots, k_l, y_1, \dots, y_w$. (The factor 2^{2^l} guarantees that the coefficients are integers.) Of course, the number

2^{2^l} Cantor$_l(k_1^D, \dots, k_l^D)$ is then a code of the set (4.1.2) in the coding generated by equation (4.1.4).

In this chapter we establish the existence of universal Diophantine equations for all n. *A priori*, it might be the case that as n is increased, the corresponding value of w would also inevitably increase. However, this turns out not to be the case: *there exists a constant* $\mathbf{m}$ *such that for every* n *there is a universal Diophantine equation* (4.1.1) *with* $l = 1$ *and* $w = \mathbf{m}$.

Once again, it is Cantor numbering that comes to our aid, enabling us to bound the number of unknowns in universal equations. Let $\mathbf{m}$ be the number of unknowns in a given universal Diophantine equation

$$U_1(a, k, y_1, \dots, y_{\mathbf{m}}) = 0 \tag{4.1.5}$$

coding one-dimensional Diophantine sets. For $n > 1$ we define the polynomial U_n by

$$U_n(a_1, \dots, a_n, k, y_1, \dots, y_{\mathbf{m}}) = U_1(2^{2^n} \operatorname{Cantor}_n(a_1, \dots, a_n), k, y_1, \dots, y_{\mathbf{m}}). \tag{4.1.6}$$

We now verify that the equation

$$U_n(a_1, \dots, a_n, k, y_1, \dots, y_{\mathbf{m}}) = 0 \tag{4.1.7}$$

is universal. Let (4.1.2) be an arbitrary Diophantine equation. Consider the one-dimensional Diophantine set $\mathfrak{D}$ defined by the equation

$$D^2(z_1, \dots, z_n, x_1, \dots, x_m) + (a - 2^{2^n} \operatorname{Cantor}_n(z_1, \dots, z_n))^2 = 0. \tag{4.1.8}$$

Since (4.1.5) is universal, there is a number $k^{\mathfrak{D}}$, a code of the set $\mathfrak{D}$, such that the equation

$$U_1(a, k^{\mathfrak{D}}, y_1, \dots, y_{\mathbf{m}}) = 0 \tag{4.1.9}$$

also defines the set $\mathfrak{D}$. It remains to note that by definition (4.1.6), the equation

$$U_n(a_1, \dots, a_n, k^{\mathfrak{D}}, y_1, \dots, y_{\mathbf{m}}) = 0 \tag{4.1.10}$$

is solvable or unsolvable precisely when equation (4.1.2) is; this implies that one can choose $k^{\mathfrak{D}}$ for the number k^D required by the definition of a universal equation.

4.2 Coding equations

As was shown in the previous section, it suffices to find a universal Diophantine equation (4.1.1) for $n = 1$. We begin by constructing a universal Diophantine equation with six code parameters b, c_{L}, c_{R}, d, e, f; i.e., we shall exhibit a polynomial U such that for every Diophantine equation

$$D(a, x_1, \ldots, x_m) = 0, \tag{4.2.1}$$

there are numbers b^D, c_{L}^D, c_{R}^D, d^D, e^D, f^D such that equation (4.2.1) is solvable for the same values of the parameter a as the equation

$$U(a, b^D, c_{\mathrm{L}}^D, c_{\mathrm{R}}^D, d^D, e^D, f^D, y_7, \ldots, y_{\mathrm{m}}) = 0. \tag{4.2.2}$$

In this chapter we are generally dealing with a given "arbitrary" Diophantine equation (4.2.1); so, in order to simplify the notation, from now on we will omit the superscript D.

The definition of *universal equation* merely requires that the sextuple

$$\langle b, c_{\mathrm{L}}, c_{\mathrm{R}}, d, e, f \rangle$$

determine the set defined by equation (4.2.1). In the universal equation that we will construct, the sextuple will determine not only this set, but also the corresponding equation itself. For this reason, the sextuple will also be called an *extended code of equation* (4.2.1).

Clearly therefore, such an extended code must contain sufficient information about the polynomial in question to determine the number of its unknowns, its degree, and its coefficients. The first two of these quantities will be present explicitly; namely, e will be equal to m, the number of unknowns in equation (4.2.1), while d can be any number greater than the total (with respect to all variables) degree of the polynomial D (without loss of generality, we suppose that the degree is greater than 0). In addition, f will be equal to

$$f = 2^{d^1} + \cdots + 2^{d^e}. \tag{4.2.3}$$

The triple $\langle d, e, f \rangle$ will be called a *format of equation* (4.2.1). Evidently, f doesn't carry any additional information. Actually, the relation $\mathrm{Format}(d, e, f)$ defined by equation (4.2.3) is Diophantine (see Exercise 4.3), but we do not need this fact.

In the previous chapter we introduced several methods for coding tuples of *natural numbers*, but, in the general case, the coefficients of the polynomial D may be arbitrary *integers*. Therefore, we need to represent the polynomial D as the difference of two polynomials with coefficients that are natural numbers:

$$D(x_0,\dots,x_m) = C_{\mathrm{L}}(x_0,\dots,x_m) - C_{\mathrm{R}}(x_0,\dots,x_m), \tag{4.2.4}$$

$$C_{\mathrm{L}}(x_0,\dots,x_m) = \sum_{i_0+\cdots+i_m<d} c_{\mathrm{L},i_0,\dots,i_m} x_0^{i_0}\dots x_m^{i_m}, \tag{4.2.5}$$

$$C_{\mathrm{R}}(x_0,\dots,x_m) = \sum_{i_0+\cdots+i_m<d} c_{\mathrm{R},i_0,\dots,i_m} x_0^{i_0}\dots x_m^{i_m}. \tag{4.2.6}$$

The numbers c_{L} and c_{R} will be ciphers to the base b of certain tuples containing information about the coefficients of the polynomials C_{L} and C_{R}, respectively. Since we will treat c_{L}, C_{L} and c_{R}, C_{R} in a similar manner, we shall often omit the subscripts L and R.

In order to define c, we first need to display the coefficients $c_{i_0,\dots,i_m}$ in a suitable linear order. In addition, these coefficients will be multiplied by certain factors, for reasons that will become clear later. Thus, we define the *cipher c of the polynomial C to the base b* as follows:

$$c = \sum_{i_0+\cdots+i_m<d} i_0!\dots i_m!\,(d-1-i_0-\cdots-i_m)!\,c_{i_0,\dots,i_m} b^{d^{m+1}-i_0d^0-\cdots-i_md^m}. \tag{4.2.7}$$

The quintuple $\langle b,c,d,e,f\rangle$ will be called a *code of the polynomial C* if in addition

$$b > d!\max\{c_{i_0,\dots,i_m}\}. \tag{4.2.8}$$

(This inequality guarantees that the coefficients $c_{i_0,\dots,i_m}$ are uniquely determined by b, c, and d.) Finally, the sextuple $\langle b,c_{\mathrm{L}},c_{\mathrm{R}},d,e,f\rangle$ will be an *extended code of equation* (4.2.1) if the quintuples $\langle b,c_{\mathrm{L}},d,e,f\rangle$ and $\langle b,c_{\mathrm{R}},d,e,f\rangle$ are codes of the polynomials C_{L} and C_{R}, respectively.

We have decided not to pause to prove that the relation Format is Diophantine; likewise, we will not take the trouble to show that the same is true of the relation *is an extended code of an equation*, written $\mathrm{ECode}(b,c_{\mathrm{L}},c_{\mathrm{R}},d,e,f)$. We shall construct a universal equation

$$U(a,b,c_{\mathrm{L}},c_{\mathrm{R}},d,e,f,y_7,\dots,y_{\mathbf{m}}) = 0 \tag{4.2.9}$$

with the following property:

$$\begin{aligned}&\mathrm{ECode}(b,c_{\mathrm{L}},c_{\mathrm{R}},d,e,f)\\ &\qquad\Longrightarrow \forall a\big[\exists y_7\dots y_{\mathbf{m}}\,[U(a,b,c_{\mathrm{L}},c_{\mathrm{R}},d,e,f,y_7,\dots,y_{\mathbf{m}})=0]\\ &\qquad\qquad\Longleftrightarrow \exists x_1\dots x_m\,[D_{b,c_{\mathrm{L}},c_{\mathrm{R}},d,e,f}(a,x_1,\dots,x_m)=0]\big],\end{aligned} \tag{4.2.10}$$

where

$$D_{b,c_{\mathrm{L}},c_{\mathrm{R}},d,e,f}(a,x_1,\ldots,x_m)=0 \tag{4.2.11}$$

is the equation with extended code $\langle b, c_{\mathrm{L}}, c_{\mathrm{R}}, d, e, f\rangle$. We shall not be concerned with the question of whether or not equation (4.2.9) is solvable in the case when $\langle b, c_{\mathrm{L}}, c_{\mathrm{R}}, d, e, f\rangle$ is not an extended code of any equation.

4.3 Coding possible solutions

In the previous section we introduced the notion of a format of an equation

$$D(a,x_1,\ldots,x_m)=0 \tag{4.3.1}$$

as a triple $\langle d,e,f\rangle$ of a special form. In this section we introduce a new coding of tuples of natural numbers $\langle x_1,\ldots,x_m\rangle$, thinking of such tuples as potential solutions of equation (4.3.1). The new coding consists of five numbers d, e, f, g, h, the first three of which constitute a format of the equation. The number g must be greater than the numbers 1, x_1, …, x_m, while h is to be defined by

$$h=x_1g^{d^1}+x_2g^{d^2}+\cdots+x_mg^{d^m}. \tag{4.3.2}$$

In other words, $\langle h,g,d^m+1\rangle$ is a positional code of the tuple

$$\langle 0,\ldots,0,x_1,0,\ldots,0,x_2,\ldots,0,\ldots,0,x_m\rangle, \tag{4.3.3}$$

in which the elements of the tuple $\langle x_1,\ldots,x_m\rangle$ are "diluted" by extra zeros, the role of which will become clear in the following section.

In Section 4.2 we decided not to pause to prove that the relation Format is Diophantine; likewise, we will not prove this for the relation *is a code of a possible solution*, written $\mathrm{SCode}(d,e,f,g,h)$. Instead, we shall easily show that another relation, SCod, is Diophantine. SCod will coincide with SCode provided that $\langle d,e,f\rangle$ is a format of an equation:

$$\mathrm{Format}(d,e,f)\Longrightarrow[\mathrm{SCod}(d,e,f,g,h)\iff\mathrm{SCode}(d,e,f,g,h)]. \tag{4.3.4}$$

The relation SCod is introduced on the basis of the following considerations: if $\langle d,e,f\rangle$ is a format of an equation, then the triple $\langle f,2,d^e+1\rangle$ is a positional code of the tuple

$$\langle 0,\ldots,0,1,0,\ldots,0,1,\ldots,0,\ldots,0,1\rangle, \tag{4.3.5}$$

in which the 1's occur in precisely the positions that are occupied by $x_1, \ldots, x_m$ in (4.3.3). Using Equal we can specify t to be equal to the cipher of tuple (4.3.5) to the base g. Then $(g-1)t$ will be the cipher to the base g of the tuple

$$\langle 0, \ldots, 0, g-1, 0, \ldots, 0, g-1, \ldots, 0, \ldots, 0, g-1 \rangle, \tag{4.3.6}$$

and it suffices to apply the relation NotGreater.

Thus, we define the relation SCod by

$$\text{SCod}(d, e, f, g, h) \iff \exists t\, [\text{Equal}(f, 2, d^e+1, t, g, d^e+1) \,\&\, \text{NotGreater}(h, (g-1)t, g)]. \tag{4.3.7}$$

It is easy to see that property (4.3.4) holds under this definition of SCod.

4.4 Computing the values of polynomials

Suppose that we have an extended code $\langle b, c_{\text{L}}, c_{\text{R}}, d, e, f\rangle$ of an equation

$$D(a, x_1, \ldots, x_m) = 0 \tag{4.4.1}$$

and a code $\langle d, e, f, g, h\rangle$ of one of its possible solutions $x_1, \ldots, x_m$. We want to determine whether $\langle d, e, f, g, h\rangle$ is indeed a code of an actual solution of equation (4.4.1) for a given value of the parameter a. An obvious way to do this is as follows: decode the coefficients of the polynomial from its extended code, decode the values of the unknowns from the code of a possible solution, and check (4.4.1). There is another method that does not require decoding. Due to our special choice of codings, we can calculate the value of the polynomial C directly from its code $\langle b, c, d, e, f\rangle$, the code $\langle d, e, f, g, h\rangle$ of a possible solution, and the value of the parameter a.

Let w be a sufficiently large number (the precise inequality will be provided later). Using Equal, we can pass from the above codes using the bases b and g, respectively, to codes $\langle w, s, d, e, f\rangle$ and $\langle d, e, f, w, t\rangle$ of the same polynomial C and the same possible solution, but now using the base w. Now consider the number

$$\begin{aligned}(1+aw+t)^{d-1} &= (1+aw^{d^0}+x_1w^{d^1}+\cdots+x_mw^{d^m})^{d-1}\\ &= \sum_{i_0+\cdots+i_m<d} \binom{d-1}{i_0 \ldots i_m} a^{i_0}x_1^{i_1}\ldots x_m^{i_m}w^{i_0d^0+\cdots+i_md^m}.\end{aligned} \tag{4.4.2}$$

Note that in this sum the exponents of w are all different, and hence if w is large enough, (4.4.2) will be the cipher to the base w of some tuple containing all the monomials constituting the polynomial C, but with different coefficients. To "touch up" the coefficients, we multiply (4.4.2) by

$$s = \sum_{i_0+\cdots+i_m<d} i_0!\dots i_m!\,(d-1-i_0-\cdots-i_m)!\,c_{i_0\dots i_m} w^{d^{m+1}-i_0 d^0-\cdots-i_m d^m} \tag{4.4.3}$$

and combine terms in which w occurs to the same power:

$$(1+aw+t)^{d-1}s = \sum_{k=0}^{2d^{m+1}-1} C_k w^k. \tag{4.4.4}$$

Here, the C_k are expressions containing $c_{i_0\dots i_m}$, a, x_1, $\dots$, x_m. It is not difficult to see that

$$\begin{aligned} C_{d^{e+1}} &= \sum_{i_0+\cdots+i_m<d} \binom{d-1}{i_0\dots i_m} i_0!\dots i_m!(d-1-i_0-\cdots-i_m)!\,c_{i_0\dots i_m} a^{i_0} x_1^{i_1}\dots x_m^{i_m} \\ &= (d-1)!\,C(a,x_1,\dots,x_m). \end{aligned} \tag{4.4.5}$$

So, if w is greater than C_0, $\dots$, $C_{2d^{m+1}-1}$, then

$$C(a,x_1,\dots,x_m) = \frac{\mathrm{Elem}((1+aw+t)^{d-1}s, w, d^{e+1}+1)}{(d-1)!}. \tag{4.4.6}$$

It remains to find an explicit lower bound for w that will guarantee (4.4.6). It is not difficult to check that the inequality

$$w > (1+a+h)^{d-1}c \tag{4.4.7}$$

will do.

Finally, we can define the relation

$$\begin{aligned} &\mathrm{Solution}(a,b,c_{\mathrm{L}},c_{\mathrm{R}},d,e,f,g,h) \iff \\ &\quad \mathrm{SCod}(d,e,f,g,h)\ \&\ \exists s_{\mathrm{L}} s_{\mathrm{R}} t w \big[w > (1+a+h)^{d-1}(c_{\mathrm{L}}+c_{\mathrm{R}}) \\ &\qquad \&\ \mathrm{Equal}(c_{\mathrm{L}}, b, d^{e+1}+1, s_{\mathrm{L}}, w, d^{e+1}+1) \\ &\qquad \&\ \mathrm{Equal}(c_{\mathrm{R}}, b, d^{e+1}+1, s_{\mathrm{R}}, w, d^{e+1}+1) \\ &\qquad \&\ \mathrm{Equal}(h, g, d^e+1, t, w, d^e+1) \\ &\qquad \&\ \mathrm{Elem}((1+aw+t)^{d-1}s_{\mathrm{L}}, w, d^{e+1}+1) \\ &\qquad\quad = \mathrm{Elem}((1+aw+t)^{d-1}s_{\mathrm{R}}, w, d^{e+1}+1)\big]. \end{aligned} \tag{4.4.8}$$

It follows from (4.4.6) that

$$\text{ECode}(b, c_{\text{L}}, c_{\text{R}}, d, e, f) \Longrightarrow$$
$$\Big[\exists gh\,[\text{SCode}(d,e,f,g,h)\ \&\ \text{Solution}(a,b,c_{\text{L}},c_{\text{R}},d,e,f,g,h)]$$
$$\Longleftrightarrow \exists x_1 \ldots x_m\,[D_{b,c_{\text{L}},c_{\text{R}},d,e,f}(a,x_1,\ldots,x_m)=0]\Big].$$

4.5 Universal Diophantine equations

Clearly, the relation $\text{Solution}(a,b,c_{\text{L}},c_{\text{R}},d,e,f,g,h)$ introduced at the end of the previous section is Diophantine, and we can construct a Diophantine equation

$$U(a,b,c_{\text{L}},c_{\text{R}},d,e,f,g,h,y_9,\ldots,y_{\mathbf{m}})=0 \tag{4.5.1}$$

defining this relation. It is easy to see that we obtain the desired universal equation (4.2.9) once we regard g and h as unknowns.

Using the technique from Section 4.1, we are now able to construct for every n a universal polynomial U_n with one code parameter and $\mathbf{m}$ unknowns. Following (4.1.4) we define the polynomial U_1 by

$$U_1(a,k,y_1,\ldots,y_{\mathbf{m}}) = U^2(a,y_1,\ldots,y_{\mathbf{m}}) + (k - 2^{2^6}\,\text{Cantor}_6(y_1,\ldots,y_6))^2. \tag{4.5.2}$$

Then if $\langle b,c_{\text{L}},c_{\text{R}},d,e,f\rangle$ is an extended code of an equation

$$D(a,x_1,\ldots,x_m)=0, \tag{4.5.3}$$

the number

$$2^{2^6}\,\text{Cantor}_6(b,c_{\text{L}},c_{\text{R}},d,e,f) \tag{4.5.4}$$

will be a *code of equation* (4.5.3) in the new coding.

Under this definition, not every natural number is a code of an equation. This "shortcoming" can easily be remedied as follows: Clearly, given a number k, we can determine whether it is representable in the form (4.5.4) with $\langle b,c_{\text{L}},c_{\text{R}},d,e,f\rangle$ being an extended code of some equation. If k is *not* representable in this form, then we shall consider that by definition k is a *code of the equation*

$$U_1(a,k,x_1,\ldots,x_{\mathbf{m}})=0 \tag{4.5.5}$$

with the single parameter a and $\mathbf{m}$ unknowns $x_1, \ldots, x_{\mathbf{m}}$. Clearly, *for any values of a and k, equation* (4.5.5) *is solvable if and only if the equation with code k is* *solvable,* regardless of whether k is a code in the original sense (4.5.4) or a code of equation (4.5.5).

According to (4.1.6), for $n > 1$ a universal polynomial U_n can be defined by

$$U_n(a_1, \dots, a_n, k, y_1, \dots, y_{\mathbf{m}}) = U_1(2^{2^n} \operatorname{Cantor}_n(a_1, \dots, a_n), k, y_1, \dots, y_{\mathbf{m}}). \tag{4.5.6}$$

Finally, we define the universal polynomial U_0 by

$$U_0(k, y_1, \dots, y_{\mathbf{m}}) = U_1(0, k, y_1, \dots, y_{\mathbf{m}}) \tag{4.5.7}$$

and consider any code of equation (4.5.3) a code of the parameter-free equation

$$D(0, x_1, \dots, x_m) = 0. \tag{4.5.8}$$

4.6 Diophantine sets with non-Diophantine complements

Now that we have universal Diophantine equations at our disposal, we can easily construct an example of a Diophantine set whose complement is not Diophantine (another example will be given in Section 6.6). The existence of such sets implies that we cannot extend our arsenal of logical tools for constructing Diophantine sets (consisting at present of $\exists$, &, and $\vee$) by adding either negation $\neg$ or universal quantification $\forall$, because the complement of the Diophantine set defined by the equation

$$D(a, x_1, \dots, x_m) = 0 \tag{4.6.1}$$

is defined both by

$$\neg \exists x_1 \dots x_m \, [D(a, x_1, \dots, x_m) = 0] \tag{4.6.2}$$

and by

$$\forall x_1 \dots x_m \, [D(a, x_1, \dots, x_m) \neq 0]. \tag{4.6.3}$$

The construction is based on the classical diagonal method. Let us consider a universal Diophantine equation

$$U_1(p, q, y_1, \dots, y_{\mathbf{m}}) = 0 \tag{4.6.4}$$

and replace both parameters by a single element parameter a:

$$U_1(a, a, y_1, \dots, y_{\mathbf{m}}) = 0. \tag{4.6.5}$$

The resulting equation defines a corresponding Diophantine set $\mathfrak{H}_1$ of natural numbers. We now show that $\overline{\mathfrak{H}}_1$, the complement of $\mathfrak{H}_1$, is not Diophantine.

Suppose the contrary; then $\overline{\mathfrak{H}}_1$ has some code k. Let us try to decide whether the equation

$$U_1(k, k, y_1, \ldots, y_{\mathbf{m}}) = 0 \tag{4.6.6}$$

has a solution. If it has, then by definition of $\mathfrak{H}_1$, $k \in \mathfrak{H}_1$. However, U_1 is universal and that implies that $k \in \overline{\mathfrak{H}}_1$. On the other hand, if equation (4.6.6) has no solution, then by definition of $\mathfrak{H}_1$, $k \notin \mathfrak{H}_1$, while U_1 being universal implies that $k \notin \overline{\mathfrak{H}}_1$. These contradictions demonstrate that $\overline{\mathfrak{H}}_1$ cannot be Diophantine.

The set $\mathfrak{H}_1$ is a somewhat artificial example. A more interesting example is furnished by the set $\mathfrak{H}_0$ defined by the equation

$$U_0(t, y_1, \ldots, y_{\mathbf{m}}) = 0. \tag{4.6.7}$$

According to this definition, $\mathfrak{H}_0$ is just the set of codes of those parameter-free Diophantine equations that have solutions. In this terminology, Hilbert's Tenth Problem is precisely the problem of providing a method for deciding whether a given number a belongs to the set $\mathfrak{H}_0$.

The polynomial U_0 was defined by equation (4.5.7). Thus, solving equation (4.6.7) reduces itself to solving equation (4.6.4) with $p = 0$, $q = t$. In order to prove that $\overline{\mathfrak{H}}_0$, the complement of $\mathfrak{H}_0$, is not Diophantine, we are going to establish an inverse relation; namely, we shall show that solving equation (4.6.4) for any parameters p and q can be reduced to solving equation (4.6.7) for an appropriate value of t. As stated, this assertion is evident, because for fixed values of p and q one can take for t any code of equation (4.6.4) considered as a parameter-free equation. What is important for us is that such a code can be given by a polynomial in p and q with integral coefficients.

So, we consider the equation

$$W_{p,q}(y_1, \ldots, y_{\mathbf{m}}) = 0 \tag{4.6.8}$$

resulting from (4.6.4) by substituting particular values for p and q. First, we construct an extended code $\langle b, c_{\mathrm{L}}, c_{\mathrm{R}}, d, e, f\rangle$ of this equation. Clearly, the values of d and e, and hence that of f, need not depend on the values of p and q. The decomposition (4.2.4), resulting in polynomials whose coefficients are natural numbers, is to be carried out for $D = U_1$ rather than for $D = W_{p,q}$, i.e., independently of the values of p and q. The result of this is that $c_{\mathrm{L},i_0,\ldots,i_{\mathbf{m}}}$ and $c_{\mathrm{R},i_0,\ldots,i_{\mathbf{m}}}$ will be polynomials in p and q with natural number coefficients. Next, we can find a polynomial expression for b satisfying (4.2.8) and polynomial expressions for the ciphers of the polynomials c_{L} and c_{R} defined by (4.2.7). Finally, by (4.5.4), a code k of equation (4.6.8) can be expressed as a polynomial in the elements of the extended code $\langle b, c_{\mathrm{L}}, c_{\mathrm{R}}, d, e, f\rangle$. Thus, $k = K(p,q)$, where K is a polynomial with integral coefficients.

Hence, we have the following relation between the sets $\mathfrak{H}_1$ and $\mathfrak{H}_0$:

$$a \in \mathfrak{H}_1 \iff K(a,a) \in \mathfrak{H}_0, \tag{4.6.9}$$

or, in terms of complements,

$$a \in \overline{\mathfrak{H}}_1 \iff K(a,a) \in \overline{\mathfrak{H}}_0. \tag{4.6.10}$$

Clearly, if $\overline{\mathfrak{H}}_0$ were Diophantine, then $\overline{\mathfrak{H}}_1$ would also be Diophantine. Since the latter is not true, we have proved that $\overline{\mathfrak{H}}_0$ is not Diophantine.

Exercises

1. In Section 4.1 we reduced the number of code parameters at the cost of an increase in the number of unknowns. Suggest a different method that will transform a universal equation (4.1.1) into another universal equation with w unknowns and a single code parameter.

2. In Section 1.4 and Exercise 1.8 we gave certain special forms for parametric Diophantine equations. Suggest corresponding special forms for universal equations.

3. Show that the relation Format is Diophantine.

4. The universal equations constructed in this chapter are rather complicated. However, once we know that there exists a universal equation

$$U(a, k, x_2, \ldots, x_m) = 0$$

with degree (with respect to all the variables) less than d, we can easily construct another universal equation with a rather regular structure. An example is the equation

$$\sum_{i_0+\cdots+i_m<d} [c_{\mathrm{L},i_0,\ldots,i_m} - c_{\mathrm{R},i_0,\ldots,i_m}] a^{i_0} k^{i_1} x_2^{i_2} \ldots x_m^{i_m} = 0,$$

in which a is an element parameter, $x_2, \ldots, x_m$ are unknowns, and the remaining variables are code parameters. We can decrease the number of code parameters (while keeping the structure regular) if we use, instead of Cantor numbering, a different polynomial that assumes each of its value only once. For example, the following equation is universal

$$\left[\sum_{i_0+\cdots+i_m<d} (c_{\mathrm{L},i_0,\ldots,i_m} - c_{\mathrm{R},i_0,\ldots,i_m})\, a^{i_0} k^{i_1} x_2^{i_2} \ldots x_m^{i_m}\right]^2$$
$$+ \left[b_{\mathrm{L}} - \sum_{i_0+\cdots+i_m<d} c_{\mathrm{L},i_0,\ldots,i_m} c^{i_0 d^0+\cdots+i_m d^m}\right]^2$$
$$+ \left[b_{\mathrm{R}} - \sum_{i_0+\cdots+i_m<d} c_{\mathrm{R},i_0,\ldots,i_m} c^{i_0 d^0+\cdots+i_m d^m}\right]^2$$
$$+ \left[c-1-\sum_{i_0+\cdots+i_m<d} (c_{\mathrm{L},i_0,\ldots,i_m} + c_{\mathrm{R},i_0,\ldots,i_m})\right]^2 = 0,$$

provided that a is considered an element parameter, b_{L}, b_{R}, c, and k code parameters, and the remaining variables are unknowns. Determine particular values of d and m for which these equations are universal.

Open question

1. Do there exist constants $\mathbf{d}$ and $\mathbf{m}$ such that for every n one can construct a universal equation with n element parameters, a single code parameter, $\mathbf{m}$ unknowns, and total degree (with respect to all the variables) equal to $\mathbf{d}$?

Commentary

The construction of a universal Diophantine equation presented in this chapter is purely number-theoretic, and in principle it could have been found in the last century. However, this didn't happen, and most likely not because of any technical difficulty, but because number theorists never suspected the possibility. As a typical example in this respect, we can mention the attitude to the paper by Davis, Putnam, and Robinson [1961] that implied the existence of a universal exponential Diophantine equation. The following was written by a reviewer of that paper, G. Kreisel [1962]:

> These results are superficially related to Hilbert's tenth Problem on (ordinary, i.e., non-exponential) Diophantine equations. The proof of the authors' results, though very elegant, does not use recondite facts in the theory of numbers nor in the theory of r.e. sets, and so it is likely that the present result is not closely connected with Hilbert's tenth Problem. Also it is not altogether plausible that all (ordinary) Diophantine problems are uniformly reducible to those in a fixed number of variables of fixed degree, which would be the case if all r.e. sets were Diophantine.

In contrast with number theory, computability theory had dealt for a long time with universal objects analogous to universal equations, in particular, with creative sets, universal Turing machines, normal forms for partial recursive functions, and so on. Martin Davis [1953] put forward a very general hypothesis (for more details see the Commentary to Chapter 5) from which it would follow that the main ideas of computability theory could be carried over to Diophantine equations and, in particular, would allow one to construct a universal Diophantine equation. Davis's hypothesis was proved in 1970 (see the Commentary to Chapter 5), and the first construction of a universal Diophantine equation was based essentially on ideas and methods of computability theory. In Chapter 6 we shall give another construction of a universal Diophantine equation based on precisely these ideas.

As soon as the mere existence of universal equations was established, it was natural to ask how small the number of unknowns in such an equation could be. The first rough estimate of 200 unknowns was stated by Matiyasevich [1971c]. Matiyasevich and Robinson [1975] improved this result to 13 unknowns by using new number-theoretic results instead of the results from computability that had originally been used. Ultimately, these techniques led to the purely number-theoretic construction introduced in this chapter. We presented it here without the many technical details that would have enabled us to reduce the number of unknowns to 9 because those details would have made it more difficult to understand the main ideas. The bound of 9 unknowns was announced by Matiyasevich [1977b] and presented with full details by Jones [1982b].

In the latter paper, Jones also exhibited the following universal system of Diophantine equations:

$$elg^2 + \alpha = (b - xy)q^2, \quad q = b^{5^{60}}, \quad \lambda + q^4 = 1 + \lambda b^5, \quad \theta + 2z = b^5,$$
$$l = u + t\theta, \quad e = y + m\theta, \quad n = q^{16},$$
$$r = \left[g + eq^3 + lq^5 + \left(2(e - z\lambda)(1 + xb^5 + g)^4 + \lambda b^5 + \lambda b^5 q^4\right)q^4\right][n^2 - n]$$
$$+ [q^3 - bl + l + \theta\lambda q^3 + (b^5 - 2)q^5][n^2 - 1],$$
$$p = 2ws^2r^2n^2, \quad p^2k^2 - k^2 = 1 = \tau^2, \quad 4(c - ksn^2)^2 + \eta = k^2,$$
$$k = r + 1 + hp - h, \quad a = (wn^2 + 1)rsn^2, \quad c = 2r + 1 + \phi,$$
$$d = bw + ca - 2c + 4a\gamma - 5\gamma, \quad d^2 = (a^2 - 1)c^2 + 1, \quad f^2 = (a^2 - 1)i^2c^4 + 1,$$
$$(d + of)^2 = \left(\left(a + f^2(d^2 - a)\right)^2 - 1\right)(2r + 1 + jc)^2 + 1.$$

Here, x is an element parameter, u, y, and z are code parameters, and the remaining 28 variables are unknowns whose values are positive integers. It should be emphasized that the proof that this system is universal was not purely number-theoretic because it was based (like Exercise 4.4) on the existence of another universal equation.

The existence of a Diophantine set with non-Diophantine complement was established by Davis [1953]. This result was obtained as a consequence of the existence of algorithmically unsolvable problems and the possibility of stating them in the full language of arithmetic. The use of a diagonal construction for proving this result takes its origin from Cantor's proof of the uncountability of the reals.

As for $\mathfrak{H}_0$, the set of codes of Diophantine equations having solutions, the fact that it is Diophantine while its complement is not is closely connected with the algorithmic unsolvability of Hilbert's Tenth Problem (for details, see Chapter 5), and this fact was established as part of the negative solution of Hilbert's Tenth Problem using logical tools. The proof in Section 4.6 seems to be the first that is purely number-theoretic.

5 Hilbert's Tenth Problem Is Unsolvable

In this chapter we discuss Turing machines, abstract computers of a simple kind, and establish a connection between them and Diophantine equations. This will enable us to give a negative solution of Hilbert's Tenth Problem.

5.1 Turing machines

The number-theoretic techniques developed in the previous chapters will permit us to tackle Hilbert's Tenth Problem from the point of view of computability theory. But first we must clarify what is to be understood by the word "process" when Hilbert speaks of "a process according to which it can be determined by a finite number of operations whether the equation is solvable in rational integers." There are many ways in which this needed clarification can be provided, but the fundamental idea is always the same. We are able to convince ourselves that we have a general method for solving problems of a particular kind if, by using this method, we can solve any individual problem of that kind without employing our creative abilities, that is, so to speak, mechanically. This last can be formalized by saying that we can construct a machine that will solve problems of the class without our intervention, i.e., automatically.

Turing machines are abstract computers. Although they can be described in purely mathematical set-theoretic terminology, we will adopt the convention of describing Turing machines as though they were physical devices.

A Turing machine has memory in the form of a *tape* divided into *cells*. The tape has a single end; to make matters definite, we assume it to be on the left. To the right, the tape is potentially infinite. This means that, in contrast with actual physical computers, a computation by a Turing machine will never lead to an abnormal termination with the diagnostic "INSUFFICIENT MEMORY." On the other hand, any particular computation will require only finitely many cells.

Each cell will either be empty or will contain a single *symbol* from a finite set of symbols $\mathrm{A} = \{\alpha_1, \ldots, \alpha_w\}$, called an *alphabet*. Different machines may have different alphabets. One of the symbols will play a special role in that it will always mark the left-most cell and appear nowhere else. We use the symbol "$\star$" for this marker. In addition, we need a symbol to denote an empty cell, and we follow tradition in using the letter "Λ" for this purpose.

Symbols on the tape are read and written by a *head*, which at each moment of discrete time scans one of the cells. The head can move along the tape to the left and to the right.

At each moment, the machine is in one of finitely many *states* that, following tradition, we will denote by $\mathrm{q}_1, \ldots, \mathrm{q}_v$. One of the states is declared to be *initial*, and we shall always suppose that this is q_1. In addition, one or more states are declared to be *final.*

The next action of a machine is totally determined by its current state and the symbol scanned by the head. In a single step, the machine can change the symbol in the cell, move the head one cell to the left or to the right, and pass into another state. The actions are defined by a set of *instructions* of the form

$$\mathrm{q}_i\alpha_j \Longrightarrow \alpha_{A(i,j)}D(i,j)\mathrm{q}_{Q(i,j)}, \tag{5.1.1}$$

where

(a) q_i, the current state, is not permitted to be final;
(b) α_j is the symbol scanned by the head, $\alpha_j \in \mathrm{A}$ unless the current cell is empty in which case $\alpha_j = \Lambda$.
(c) $\alpha_{A(i,j)}$ is the symbol to be written, $\alpha_{A(i,j)} \in \mathrm{A}$ (the case $\alpha_{A(i,j)} = \alpha_j$ is not excluded);
(d) $D(i,j)$ represents motion by the head—there are three options:

(i) L, the head moves one cell to the left,
(ii) R, the head moves one cell to the right,
(iii) S, the head stands still;

(e) $\mathrm{q}_{Q(i,j)}$ is the new state (the case $\mathrm{q}_{Q(i,j)} = \mathrm{q}_i$ is not excluded).

We refer to the part of an instruction preceding the $\Longrightarrow$ as its left-hand side, and the part following the $\Longrightarrow$ as its right-hand side. Following our convention, if $\alpha_j = \star$, then $\alpha_{A(i,j)} = \star$ and $D(i,j) \neq \mathrm{L}$; on the other hand, if $\alpha_j \neq \star$, then $\alpha_{A(i,j)} \neq \star$. For each pair consisting of a non-final state and an element of $\mathrm{A} \cup \{\Lambda\}$, there must be *exactly* one instruction with the corresponding left-hand side.

To begin with, some initial segment of the tape is occupied by symbols from A with no gaps, the rest of the potentially infinite tape is empty, the head scans one of the cells, and the machine is in the initial state q_1. Then the work of the machine proceeds step by step in accordance with the instructions. The work will stop after an instruction in which $\mathrm{q}_{Q(i,j)}$ is a final state is executed. It is possible that the machine will never reach a final state and that the work will continue indefinitely.

The input information is determined by the initial content of the tape and the position of the head on the tape. The result of the computation, the output, is determined when the work stops by the content of the tape, the position of the head on the tape, and the final state. The acts of interpreting the input and output information are events external to the Turing machine. For example, natural numbers could be written on the tape in binary notation using two symbols α and β, α being zero and β one, or vice versa; it is clear that in these two cases the same machine would, in general, compute two different functions.

5.2 Composition of machines

We need to establish the existence of a number of machines with particular properties. In principle, we could exhibit the full system of instructions for each of these machines, but this would turn out to be cumbersome and impracticable. Instead, we shall explicitly provide instructions only for several of the simplest machines and describe two methods for constructing more complex machines from given machines. This will enable us to obtain machines with the required properties.

All the machines that we are going to construct will have the same alphabet,

$$\{\star, 0, 1, 2, 3, \lambda\}. \tag{5.2.1}$$

There will be two final states, q_2 and q_3, and we shall interpret reaching q_2 as the answer "YES" and reaching q_3 as the answer "NO." The cells containing the symbol "λ" will play the role of *proxies* for empty cells, in the following sense: only empty cells and cells containing the symbol "λ" may be situated to the right of a cell containing the symbol "λ", and for any state q_i, the instructions with left-hand sides $\mathrm{q}_i\lambda$ and $\mathrm{q}_i\Lambda$ will have identical right-hand sides.

The first method for constructing a new Turing machine M from two given machines M_1 and M_2 is as follows:

(a) *In all instructions of machine* M_1, *the final state* q_2 *is replaced by* q_{v+1}, *where* v *is the number of states of machine* M_1 (it should be recalled that final states can occur only in the right-hand sides of instructions).

(b) *In all instructions of machine* M_2, *every non-final state* q_i *is replaced by* q_{v+i} (in particular, q_1 is replaced by q_{v+1}).

(c) *The set of instructions of the new machine* M *consists of the instructions of both of the given machines, modified as described above.*

The action of machine M clearly consists of the consecutive execution of the actions of machines M_1 and M_2 *as originally constituted*, provided that machine M_1 halted

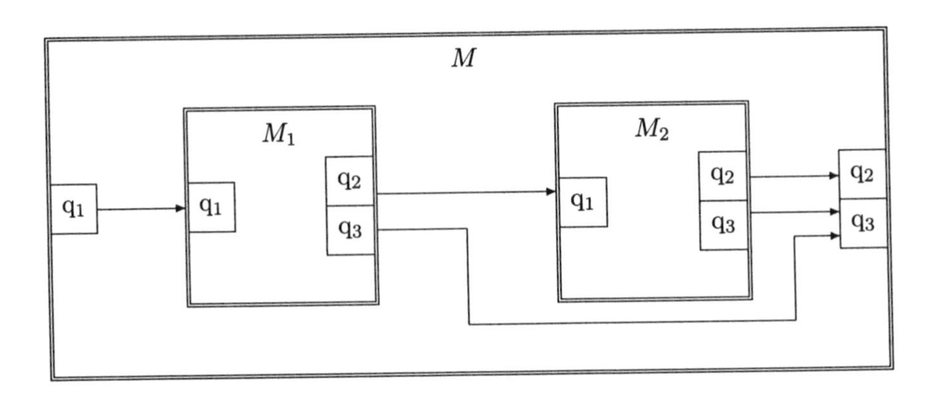

Figure 5.1. The first method of composition

in state q_2. To denote machine M, we shall use any one of the three notations

$$M_1; M_2$$

$$M_1 \textbf{ and } M_2$$

or

$$\textbf{if } M_1 \textbf{ then } M_2$$

depending on which seems most natural in a given context.

It easy to see that the composition of Turing machines described above is an associative operation, so notation like $M_1; M_2; M_3$ is unambiguous.

The second method for constructing a new Turing machine M from two given machines M_1 and M_2 is as follows:

(a) *In all instructions of machine M_1, the final state q_2 is replaced by q_{v+1}, where v is the number of states of machine M_1, and the final state q_3 is replaced by q_2.*
(b) *In all instructions of machine M_2, every non-final state q_i is replaced by q_{v+i}, and the final state q_2 is replaced by q_1.*
(c) *The set of instructions of the new machine M consists of the instructions of both of the given machines, modified as described above.*

The Turing machine constructed in this way will be denoted by

$$\textbf{while } M_1 \textbf{ do } M_2 \textbf{ od}$$

The action of this machine consists in performing in turn the actions of machines

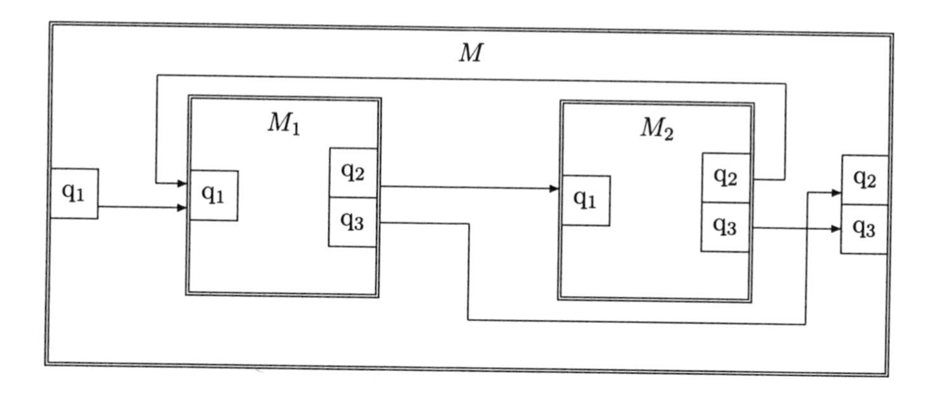

Figure 5.2. The second method of composition

M_1 and M_2 *as originally constituted* until one of them enters the final state q_3.

The notation introduced above resembles a primitive programming language. In fact, every such "program" denotes a particular Turing machine.

5.3 Basis machines

In this section we construct a number of specific Turing machines and describe their actions. While Turing machines can be thought of as working with strings of characters, we are most interested in dealing with numbers. For this we adopt a number of conventions for representing natural numbers on the tape. First of all, we shall use unary notation, namely, *the representation of a number m will* occupy $m+1$ consecutive cells. The leftmost of these will contain the symbol "0", while the remainder will contain the symbol "1". Naturally, the cell adjacent to the rightmost cell of such a representation cannot contain the symbol "1".

A representation of the tuple

$$\langle a_1, \ldots, a_n \rangle \tag{5.3.1}$$

will consist of the representations of the numbers $a_1, \ldots, a_n$ arranged one after another without gaps, beginning with the second cell of the tape. (According to our agreement, the first cell always contains the end-marker "$\star$".) The other cells may either be empty or, according to our convention from Section 5.2, contain the symbol "λ". The representation with no cells containing λ will be called *canonical.*

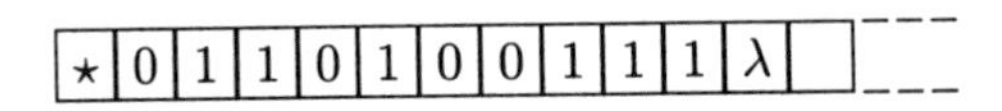

Figure 5.3. A representation of the tuple $\langle 2, 1, 0, 3\rangle$

We begin with a number of extremely simple machines that will be presented explicitly.

Machine LEFT with instructions

$$\begin{aligned}
\mathrm{q}_1\star &\Longrightarrow \star\mathrm{S}\mathrm{q}_2\\
\mathrm{q}_1 0 &\Longrightarrow 0\mathrm{L}\mathrm{q}_2\\
\mathrm{q}_1 1 &\Longrightarrow 1\mathrm{L}\mathrm{q}_2\\
\mathrm{q}_1 2 &\Longrightarrow 2\mathrm{L}\mathrm{q}_2\\
\mathrm{q}_1 3 &\Longrightarrow 3\mathrm{L}\mathrm{q}_2\\
\mathrm{q}_1\lambda &\Longrightarrow \lambda\mathrm{L}\mathrm{q}_2\\
\mathrm{q}_1\Lambda &\Longrightarrow \lambda\mathrm{L}\mathrm{q}_2
\end{aligned} \tag{5.3.2}$$

moves the head one cell to the left unless it was already scanning the leftmost cell marked by the symbol "$\star$".

Machine RIGHT with instructions

$$\begin{aligned}
\mathrm{q}_1\star &\Longrightarrow \star\mathrm{R}\mathrm{q}_2\\
\mathrm{q}_1 0 &\Longrightarrow 0\mathrm{R}\mathrm{q}_2\\
\mathrm{q}_1 1 &\Longrightarrow 1\mathrm{R}\mathrm{q}_2\\
\mathrm{q}_1 2 &\Longrightarrow 2\mathrm{R}\mathrm{q}_2\\
\mathrm{q}_1 3 &\Longrightarrow 3\mathrm{R}\mathrm{q}_2\\
\mathrm{q}_1\lambda &\Longrightarrow \lambda\mathrm{S}\mathrm{q}_2\\
\mathrm{q}_1\Lambda &\Longrightarrow \lambda\mathrm{S}\mathrm{q}_2
\end{aligned} \tag{5.3.3}$$

moves the head one cell to the right unless it was scanning an empty cell or a cell containing the symbol "λ".

Machine WRITE(0) with instructions

$$\begin{aligned}
q_1\star &\Longrightarrow \star S q_2\\
q_1 0 &\Longrightarrow 0 S q_2\\
q_1 1 &\Longrightarrow 0 S q_2\\
q_1 2 &\Longrightarrow 0 S q_2\\
q_1 3 &\Longrightarrow 0 S q_2\\
q_1 \lambda &\Longrightarrow 0 S q_2\\
q_1 \Lambda &\Longrightarrow 0 S q_2
\end{aligned} \tag{5.3.4}$$

writes the symbol "0" to the cell scanned by the head unless it is the leftmost cell containing the marker "$\star$". Similar actions are performed by the machines WRITE(1), WRITE(2), WRITE(3), and WRITE(λ), the instructions for which can be obtained from (5.3.4) by replacing the symbol "0" in the right-hand sides of the instructions by "1", "2", "3", and "λ", respectively.

Machine READ(0) with instructions

$$\begin{aligned}
q_1\star &\Longrightarrow \star S q_3\\
q_1 0 &\Longrightarrow 0 S q_2\\
q_1 1 &\Longrightarrow 1 S q_3\\
q_1 2 &\Longrightarrow 2 S q_3\\
q_1 3 &\Longrightarrow 3 S q_3\\
q_1 \lambda &\Longrightarrow \lambda S q_3\\
q_1 \Lambda &\Longrightarrow \lambda S q_3
\end{aligned} \tag{5.3.5}$$

determines whether the cell scanned by the head contains the symbol "0" or not and then halts in state q_2 or q_3, accordingly; by our convention these correspond respectively to "YES" or "NO" answers. In a similar manner, machines READ(1), READ(2), READ(3) and READ($\star$) determine the presence of symbols "1", "2", "3", and "$\star$", respectively. Machine READ(λ) with instructions

$$\begin{aligned} q_1\star &\Longrightarrow \star S q_3 \\ q_1 0 &\Longrightarrow 0 S q_3 \\ q_1 1 &\Longrightarrow 1 S q_3 \\ q_1 2 &\Longrightarrow 2 S q_3 \\ q_1 3 &\Longrightarrow 3 S q_3 \\ q_1 \lambda &\Longrightarrow \lambda S q_2 \\ q_1 \Lambda &\Longrightarrow \lambda S q_2 \end{aligned} \tag{5.3.6}$$

acts a bit differently because we have decided to consider cells containing the symbol "λ" as proxies for empty cells.

We also need the following two machines that make almost no changes on the tape. Machine STOP with instructions

$$\begin{aligned} q_1\star &\Longrightarrow \star S q_3 \\ q_1 0 &\Longrightarrow 0 S q_3 \\ q_1 1 &\Longrightarrow 1 S q_3 \\ q_1 2 &\Longrightarrow 2 S q_3 \\ q_1 3 &\Longrightarrow 3 S q_3 \\ q_1 \lambda &\Longrightarrow \lambda S q_3 \\ q_1 \Lambda &\Longrightarrow \lambda S q_3 \end{aligned} \tag{5.3.7}$$

goes directly into the final state q_3 from the state q_1, without moving the head. On the other hand, machine NEVERSTOP with instructions

$$\begin{aligned} q_1\star &\Longrightarrow \star S q_1 \\ q_1 0 &\Longrightarrow 0 S q_1 \\ q_1 1 &\Longrightarrow 1 S q_1 \\ q_1 2 &\Longrightarrow 2 S q_1 \\ q_1 3 &\Longrightarrow 3 S q_1 \\ q_1 \lambda &\Longrightarrow \lambda S q_1 \\ q_1 \Lambda &\Longrightarrow \lambda S q_1 \end{aligned} \tag{5.3.8}$$

never changes its state and hence never halts.

The action of machine

$$\text{READNOT}(0) = \textbf{while } \text{READ}(0) \textbf{ do } \text{STOP} \textbf{ od}$$

is the opposite of the action of machine READ(0); namely, it recognizes the absence of the symbol "0" in the cell observed by the head. Machines READNOT(2), READNOT(3), READNOT($\star$), and READNOT(λ) are obtained by replacing READ(0) by READ(2), READ(3), READ($\star$), and READ(λ), respectively.

Machine

$$\text{STAR} = \textbf{while } \text{READNOT}(\star) \textbf{ do } \text{LEFT} \textbf{ od}$$

puts the head into the leftmost cell (marked by "$\star$").

Machine

$$\text{VACANT} = \text{STAR}; \textbf{while } \text{READNOT}(\lambda) \textbf{ do } \text{RIGHT} \textbf{ od}$$

puts the head into the leftmost cell containing the symbol "λ" if such exists; otherwise it puts the head into the leftmost empty cell.

Machine

$$\text{JUMP} = \textbf{while } \text{READNOT}(0) \textbf{ do } \text{RIGHT} \textbf{ od}$$

moves the head to the right until the first cell containing the symbol "0" is reached; if all cells containing "0" are to the left of the head, then the machine will never halt.

A sequence of machines FIND is defined by the recurrent relations

$$\text{FIND}(1) = \text{STAR}; \text{JUMP}$$
$$\text{FIND}(k+1) = \text{FIND}(k); \text{RIGHT}; \text{JUMP}$$

Clearly, machine FIND(k) moves the head into the cell containing the symbol "0" that begins the representation of the element a_k from tuple (5.3.1).

Machine

$$\text{LAST} = \text{VACANT}; \textbf{while } \text{READNOT}(0) \textbf{ do } \text{LEFT} \textbf{ od}$$

acts on tuple (5.3.1) in the same way as machine FIND(n); namely, it moves the head into the cell containing the symbol "0" that begins the representation of the last element a_n. Thus, in an appropriate situation, the single machine LAST will perform the action of any one of the infinite sequence of machines FIND(k).

Machine

$$\text{NEW} = \text{VACANT}; \text{WRITE}(0)$$

transforms tuple (5.3.1) into tuple $\langle a_1, \ldots, a_n, 0 \rangle$.

Machine

$$\text{INC} = \text{VACANT}; \text{WRITE}(1)$$

transforms tuple (5.3.1) into the tuple

$$\langle a_1, \ldots, a_{n-1}, a_n + 1 \rangle. \tag{5.3.9}$$

On the other hand, machine

$$\text{DEC} = \text{VACANT}; \text{LEFT}; \textbf{if } \text{READ}(1) \textbf{ then } \text{WRITE}(\lambda)$$

transforms tuple (5.3.9) into tuple (5.3.1). When applied to a tuple with zero as its last element, it does not alter anything on the tape, but it halts in the state q_3, indicating that the last element has not been decreased.

Machine

$$\begin{aligned} \text{DELETE} = \text{VACANT};& \\ &\textbf{while } \text{READNOT}(0) \textbf{ do } \text{WRITE}(\lambda); \text{LEFT } \textbf{od}; \\ &\text{WRITE}(\lambda) \end{aligned}$$

truncates tuple (5.3.1), yielding the tuple $\langle a_1, \ldots, a_{n-1} \rangle$.

Machine

$$\text{MARK}(2) = \textbf{while } \text{RIGHT}; \text{READ}(1) \textbf{ do } \text{WRITE}(2) \textbf{ od}$$

replaces consecutive occurrences of the symbol "1" by the symbol "2". The machine MARK(2) is typically used after one of the machines FIND; thus, machine FIND(k); MARK(2) selects the element a_k from the representation of tuple (5.3.1).

Machine

$$\text{MARK}(3) = \textbf{while } \text{RIGHT}; \text{READ}(1) \textbf{ do } \text{WRITE}(3) \textbf{ od}$$

behaves in a similar manner.

Machine

$$\begin{aligned} \text{THEREIS}(2) = \text{STAR};& \\ &\textbf{while } \text{READNOT}(2) \textbf{ do} \\ &\quad \textbf{if } \text{READNOT}(\lambda) \textbf{ then } \text{RIGHT} \\ &\textbf{od} \end{aligned}$$

determines whether the symbol "2" occurs anywhere on the tape; if so, it halts in state q_2; otherwise it halts in state q_3. Our decision to construct all our machines in such a way that a cell containing the symbol "2" cannot be situated to the right of a cell containing the symbol "λ" is essential if machine THEREIS(2) is to perform correctly.

Machine

$$\text{THEREWAS}(2) = \textbf{if } \text{THEREIS}(2) \textbf{ then } \text{WRITE}(1)$$

not only determines the presence of the symbol "2" but also, having found a cell containing this symbol, replaces it by the symbol "1".

Machines THEREIS(3) and THEREWAS(3) are defined and act analogously to machines THEREIS(2) and THEREWAS(2), respectively.

Machines THEREWAS(2) and THEREWAS(3) restore the symbol "1" to just one cell. In contrast, machine

$$\begin{aligned}\text{RESTORE} = {} & \textbf{while } \text{THEREIS}(2) \textbf{ do } \text{THEREWAS}(2) \textbf{ od};\\ & \textbf{while } \text{THEREIS}(3) \textbf{ do } \text{THEREWAS}(3) \textbf{ od}\end{aligned}$$

restores all symbols "1" that machines MARK(2) and MARK(3) had earlier replaced by "2" or "3".

Machine

$$\begin{aligned}\text{APPEND}(k) = {} & \text{FIND}(k); \text{MARK}(2);\\ & \textbf{while } \text{THEREWAS}(2) \textbf{ do } \text{INC} \textbf{ od}\end{aligned}$$

transforms tuple (5.3.1) into $\langle a_1, \ldots, a_{n-1}, a_n + a_k \rangle$. Machine

$$\text{COPY}(k) = \text{NEW}; \text{APPEND}(k)$$

transforms tuple (5.3.1) into $\langle a_1, \ldots, a_n, a_k \rangle$. Finally, machine

$$\text{ADD}(k, l) = \text{COPY}(k); \text{APPEND}(l)$$

transforms tuple (5.3.1) into $\langle a_1, \ldots, a_n, a_k + a_l \rangle$.

For $k \neq l$, machine

$$\begin{aligned}\text{MULT}(k, l) = {} & \text{NEW}; \text{FIND}(k); \text{MARK}(3);\\ & \textbf{while } \text{THEREWAS}(3) \textbf{ do } \text{APPEND}(l) \textbf{ od}\end{aligned}$$

transforms tuple (5.3.1) into $\langle a_1, \ldots, a_n, a_k a_l \rangle$. To obtain tuple $\langle a_1, \ldots, a_n, a_k^2 \rangle$, one can use machine

```
MULT(k, k) = COPY(k); LAST; MARK(3);
             while THEREWAS(3) do
                  while THEREWAS(3) do APPEND(k) od
             od
```

For $k \neq l$, machine

```
NOTGREATER(k, l) = FIND(k); MARK(2); FIND(l); MARK(3);
                   while THEREIS(2) and THEREIS(3) do
                        THEREWAS(2); THEREWAS(3)
                   od;
                   while THEREIS(2) do
                        RESTORE; STOP
                   od;
                   RESTORE
```

compares a_k and a_l and stops in state q_2 or q_3 depending on which of the two inequalities $a_k \leq a_l$ or $a_k > a_l$ holds. Similarly, machines

```
EQUAL(k, l) = NOTGREATER(k, l) and NOTGREATER(l, k)
```

and

```
NOTEQUAL(k, l) = while EQUAL(k, l) do STOP od
```

determine whether a_k and a_l are equal.

Machine

```
NEXT = LAST; WRITE(1); RIGHT;
       while READ(λ) do WRITE(1); LAST; RIGHT od;
       WRITE(0)
```

transforms tuple (5.3.1) into the tuple $\langle a_1, \ldots, a_{n-2}, b, c\rangle$, where $\langle b, c\rangle$ is the pair that follows immediately after the pair $\langle a_{n-1}, a_n\rangle$ in Cantor's list of all pairs (3.1.1). Finally, machine

```
DECODE = LAST; MARK(2); NEW; NEW;
              while THEREWAS(2) do NEXT od
```

transforms tuple (5.3.1) into $\langle a_1, \ldots, a_n, b, c\rangle$, where $\langle b, c\rangle$ is the pair with Cantor number a_n.

5.4 Turing machines can recognize Diophantine sets

In this section we establish our first connection between Diophantine sets and Turing machines. Namely, we show that *given any parametric Diophantine equation*

$$D(a_1, \ldots, a_n, x_1, \ldots, x_{m+1}) = 0, \tag{5.4.1}$$

one can construct a Turing machine M that will eventually halt, beginning with a representation of the tuple

$$\langle a_1, \ldots, a_n\rangle, \tag{5.4.2}$$

if and only if equation (5.4.1) *is solvable in the unknowns $x_1, \ldots, x_{m+1}$.*

We begin by constructing an auxiliary machine M_1 that, given a tuple

$$\langle a_1, \ldots, a_n, y_0\rangle, \tag{5.4.3}$$

determines whether y_0 is the Cantor number of a tuple $\langle x_1, \ldots, x_{m+1}\rangle$ that does not satisfy equation (5.4.1). This machine M_1 is itself constructed by concatenating a number of other machines using the operation ";". First, we use m copies of the machine DECODE, so that the tuple (5.4.3) is transformed into the tuple

$$\langle a_1, \ldots, a_n, y_0, x_1, y_1, \ldots, x_m, y_m\rangle, \tag{5.4.4}$$

where

$$y_{k-1} = \text{Cantor}(x_k, y_k), \tag{5.4.5}$$

and hence

$$y_0 = \text{Cantor}_{m+1}(x_1, \ldots, x_m, y_m). \tag{5.4.6}$$

Equation (5.4.1) can be rewritten in the form

$$C_{\text{L}}(a_1, \ldots, a_n, x_1, \ldots, x_m, y_m) = C_{\text{R}}(a_1, \ldots, a_n, x_1, \ldots, x_m, y_m), \tag{5.4.7}$$

where C_{L} and C_{R} are polynomials whose coefficients are natural numbers and the unknown x_{m+1} has been replaced by y_m. The computation of the values of C_{L}

and C_{R} can be represented as a sequence of elementary operations

$$\begin{aligned} z_1 &= \alpha_1 \circ_1 \beta_1, \\ &\vdots \\ z_k &= \alpha_k \circ_k \beta_k, \end{aligned} \tag{5.4.8}$$

where each of α_i and β_i, $i = 1, \ldots, k$, is one of the quantities 1, a_1, ..., a_n, x_1, ..., x_m, y_m, z_1, ..., z_{k-1}, where each $\circ_i$ is one of the symbols + or $\cdot$, and where the values of C_{L} and C_{R} occur among the quantities z_1, ..., z_k. It is clear that the machines NEW, INC, ADD, and MULT can be combined to transform tuple (5.4.4) into the tuple

$$\langle a_1, \ldots, a_n, y_0, x_1, y_1, \ldots, x_m, y_m, 1, z_1, \ldots, z_k \rangle. \tag{5.4.9}$$

To complete the construction of machine M_1, it suffices to apply the machine NOTEQUAL with appropriate arguments. Thus, machine M_1 will halt in state q_2 if y_0 from tuple (5.4.2) is *not* the Cantor number of a solution of equation (5.4.1), and in state q_3 if it is.

Let M_2 denote the machine DELETE; DELETE; ... DELETE, where the machine DELETE is iterated $2m + k + 1$ times. Clearly, M_2 transforms tuple (5.4.9) into the original tuple (5.4.3).

Now we can define the machine M as

$$M = \text{NEW}; \textbf{while } M_1 \textbf{ do } M_2; \text{INC } \textbf{od} \tag{5.4.10}$$

which behaves as follows: It begins (using machine NEW) by transforming tuple (5.4.2) into the tuple

$$\langle a_1, \ldots, a_n, 0 \rangle. \tag{5.4.11}$$

If machine M_1 determines that 0 is the Cantor number of a solution of equation (5.4.1), then machine M immediately halts. Otherwise, machines M_2 and INC act to construct the tuple

$$\langle a_1, \ldots, a_n, 1 \rangle, \tag{5.4.12}$$

and machine M_1 proceeds to determine whether 1 is the Cantor number of a solution, and so on. If equation (5.4.1) has a solution, then machine M will eventually halt; if equation (5.4.1) has no solution, then machine M will continue forever.

5.5 Diophantine simulation of Turing machines

In the previous section we showed that Diophantine equations are "semidecidable"; i.e., if an equation has a solution, then this fact can be revealed purely mechanically. This "semidecidability" is an intuitively obvious property, and the formal proof given in the previous section does not contribute much to our knowledge about Diophantine equations. Instead, the proof gives us evidence that Turing machines are sufficiently powerful, in spite of their primitive set of instructions.

In fact, the main role of the previous section was to motivate the introduction of a new notion. Namely, we shall say that a set $\mathfrak{M}$ of n-tuples of natural numbers is *Turing semidecidable* if there is a Turing machine M that, beginning in state q_1, with a tape containing the canonical representation of the tuple $\langle a_1, \ldots, a_n\rangle$, and with its head scanning the leftmost cell on the tape, will eventually halt if and only if $\langle a_1, \ldots, a_n\rangle \in \mathfrak{M}$. In this case, we say that M *semidecides* $\mathfrak{M}$.

In the previous section we proved that every Diophantine set is Turing semidecidable. The aim of the present section is to establish the converse implication: *every Turing semidecidable set is Diophantine.*

Let M be a Turing machine that semidecides a set $\mathfrak{M}$ consisting of n-tuples of natural numbers. Let

$$\{\alpha_1, \ldots, \alpha_w\} \tag{5.5.1}$$

be the alphabet of M. As the machine M carries out its operations, at each moment the symbols of (5.5.1) occupy only a finite initial segment of the tape of length, say, l, and we can therefore represent the tape by the tuple

$$\langle s_1, s_2, \ldots, s_m, \ldots, s_{l-1}, s_l\rangle \tag{5.5.2}$$

consisting of the subscripts of the symbols occurring in the cells (in particular, s_1 is always equal to the subscript of the symbol "$\star$" in (5.5.1)).

The current state q_i and the position of the head can be represented by a tuple of the same length,

$$\langle 0, \ldots, 0, i, 0, \ldots, 0\rangle, \tag{5.5.3}$$

in which all elements but one are zero; the only non-zero element is equal to the subscript of the state, and its position corresponds to the position of the head.

The triple consisting of the current contents of the tape, the state, and the position of the head will be called the *configuration*. Clearly, tuples (5.5.2) and (5.5.3) uniquely determine the configuration. To represent these tuples, we shall use positional coding with a fixed base β that must be no less than 3, greater than v, the number of states of machine M, and greater than w, the number of symbols in the alphabet (5.5.1). The pair $\langle p, t\rangle$ is called a *configuration code* if p and t are ciphers of tuples (5.5.3) and (5.5.2), respectively, to the base β. A configuration code does not specify the length of tuples (5.5.2) and (5.5.3), but this quantity is of no significance because tuple (5.5.3) has only one non-zero element, while zero elements of tuple (5.5.2) will be interpreted as empty cells. Thus, the configuration is uniquely determined by a configuration code, and our use of the ciphers of tuples (5.5.2) and (5.5.3) (rather than their codes) corresponds to the potential infinity of the tape.

Using this terminology, our first goal will be to construct a Diophantine equation

$$D(p, t, x_1, \ldots, x_m) = 0 \tag{5.5.4}$$

such that *if* $\langle p, t\rangle$ *is the code of a configuration, then equation* (5.5.4) *is solvable in* $x_1, \ldots, x_m$ *if and only if machine* M*, beginning in this configuration, eventually halts.* We shall not be concerned about whether or not equation (5.5.4) is solvable when $\langle p, t\rangle$ is not a configuration code.

Naturally, we begin by simulating a single step of a Turing machine. Thus, let machine M proceed directly from the configuration with code $\langle p, t\rangle$ to the configuration with code

$$\langle \mathrm{NextP}(p, t), \mathrm{NextT}(p, t)\rangle.$$

We will now verify that NextP and NextT are Diophantine functions.

We need to make this statement more precise. So far we have defined the functions NextP and NextT only when $\langle p, t\rangle$ is a code of a configuration with a non-final state. We now set

$$\mathrm{NextP}(p, t) = 0, \tag{5.5.5}$$

$$\mathrm{NextT}(p, t) = t \tag{5.5.6}$$

if $\langle p, t\rangle$ is a configuration code with a final state. We shall understand the statement that NextP and NextT are Diophantine as meaning that there exist Diophantine functions that are equal to NextP and NextT when $\langle p, t\rangle$ is a configuration code, but which may have any value, or even be undefined, when it is not.

Clearly, the functions NextP and NextT are determined by the functions A, D, and Q from the set (5.1.1) of instructions of machine M. Following (5.5.5) and (5.5.6), we extend these functions, setting $A(i,j) = j$, $D(i,j) = S$, and $Q(i,j) = 0$ when q_i is a final state.

The fact that the function NextT is Diophantine is almost obvious. Indeed, each element of the tuple with cipher $\mathrm{NextT}(p,t)$ is uniquely determined by the elements in the same position from the tuples with ciphers p and t. Let us define the function A by

$$\mathrm{A}(i,j) = \begin{cases} A(i,j) & \text{if } 0 < i \leq v,\ 0 \leq j \leq w, \\ j & \text{otherwise.} \end{cases} \tag{5.5.7}$$

(Recall that v is the number of states of the machine M and w is the number of symbols in its alphabet.) Now we can define NextT by

$$t' = \mathrm{NextT}(p,t) \iff \exists w\,[t' = \mathrm{A}[\beta](p,t,w)], \tag{5.5.8}$$

where $\mathrm{A}[\beta]$ is the extension of A to tuples introduced in Section 3.6.

The fact that the function NextP is also Diophantine is a bit less evident. Observe that the kth element of the tuple with cipher $\mathrm{NextP}(p,t)$ is uniquely determined by the $(k-1)$th, kth, and $(k+1)$th elements of the tuples with ciphers p and t. The technique introduced in Section 3.6 only allows us to extend functions element-wise, but it is not difficult to overcome this obstacle. Put

$$p^{\mathrm{R}} = p\beta, \qquad p^{\mathrm{L}} = p \operatorname{div} \beta, \tag{5.5.9}$$

$$t^{\mathrm{R}} = t\beta, \qquad t^{\mathrm{L}} = t \operatorname{div} \beta. \tag{5.5.10}$$

Clearly, if t is the cipher of tuple (5.5.2), then t^{R} and t^{L} are the ciphers of the tuples

$$\langle 0, s_1, \dots, s_{m-1}, \dots, s_{l-1}, s_l \rangle \tag{5.5.11}$$

and

$$\langle s_2, s_3, \dots, s_{m+1}, \dots, s_l, 0 \rangle, \tag{5.5.12}$$

respectively (it is the mth elements that are shown explicitly in tuples (5.5.2), (5.5.11), and (5.5.12)). Similarly, if p is the cipher of tuple (5.5.3), then p^{R} and p^{L} are the ciphers of the tuples

$$\langle 0, \dots, 0, 0, i, \dots, 0 \rangle \tag{5.5.13}$$

and

$$\langle 0, \dots, i, 0, 0, \dots, 0 \rangle, \tag{5.5.14}$$

where the non-zero element in (5.5.3) has been shifted by one position to the right or the left, respectively.

Now every element of the tuple with cipher $\mathrm{NextP}(p,t)$ is uniquely determined by the elements in the same position in the tuples with ciphers p^{L}, p, p^{R}, t^{L}, t, t^{R} in a manner determined by the functions D and Q. To be more precise, let us introduce the function DQ by

$$\mathrm{DQ}(i^{\mathrm{L}},i,i^{\mathrm{R}},j^{\mathrm{L}},j,j^{\mathrm{R}}) = \begin{cases} Q(i^{\mathrm{L}},j^{\mathrm{L}}) & \text{if } i^{\mathrm{L}}>0,\ i=i^{\mathrm{R}}=0, \text{ and } D(i^{\mathrm{L}},j^{\mathrm{L}})=\mathrm{L}, \\ Q(i,j) & \text{if } i^{\mathrm{L}}=0,\ i>0,\ i^{\mathrm{R}}=0, \text{ and } D(i,j)=\mathrm{S}, \\ Q(i^{\mathrm{R}},j^{\mathrm{R}}) & \text{if } i^{\mathrm{L}}=i=0,\ i^{\mathrm{R}}>0, \text{ and } D(i^{\mathrm{R}},j^{\mathrm{R}})=\mathrm{R}, \\ 0 & \text{otherwise} \end{cases} \tag{5.5.15}$$

(the "otherwise" also includes the cases when some of the numbers i^{L}, i, or i^{R} are greater than v and when some of the numbers j^{L}, j, or j^{R} are greater than w, because Q and D are not defined in those cases). Now we are in position to give a Diophantine representation for NextP:

$$p' = \mathrm{NextP}(p,t) \iff \exists w\,[p' = \mathrm{DQ}[\beta](p\beta, p, p \operatorname{div} \beta, t\beta, t, t \operatorname{div} \beta, w)]. \tag{5.5.16}$$

The functions NextP and NextT simulate a single step of our Turing machine. Multiple steps are similarly simulated by the 3-argument functions AfterP and AfterT defined as iterations of functions NextP and NextT:

$$\begin{aligned} \mathrm{AfterP}(0,p,t) &= p, \\ \mathrm{AfterT}(0,p,t) &= t, \\ \mathrm{AfterP}(k+1,p,t) &= \mathrm{NextP}(\mathrm{AfterP}(k,p,t),\mathrm{AfterT}(k,p,t)), \\ \mathrm{AfterT}(k+1,p,t) &= \mathrm{NextT}(\mathrm{AfterP}(k,p,t),\mathrm{AfterT}(k,p,t)). \end{aligned} \tag{5.5.17}$$

Clearly, if machine M passes from configuration $\langle p,t\rangle$ to configuration $\langle p',t'\rangle$ in k steps, then $p' = \mathrm{AfterP}(k,p,t)$ and $t' = \mathrm{AfterT}(k,p,t)$. Next we will show that the functions AfterP and AfterT are also Diophantine.

Let us consider all of the intermediate configurations

$$\langle p_0,t_0\rangle,\ldots,\langle p_k,t_k\rangle, \tag{5.5.18}$$

where

$$\langle p_0, t_0 \rangle = \langle p, t \rangle, \tag{5.5.19}$$

$$\langle p_{i+1}, t_{i+1} \rangle = \langle \mathrm{NextP}(p_i, t_i), \mathrm{NextT}(p_i, t_i) \rangle, \tag{5.5.20}$$

$$\langle p', t' \rangle = \langle p_k, t_k \rangle. \tag{5.5.21}$$

Let the number l be so large that

$$p < \beta^{l-k-2}, \qquad t < \beta^{l-k-2}. \tag{5.5.22}$$

This implies that in the configuration coded by $\langle p, t \rangle$, no more than $l - k - 2$ initial cells are occupied by symbols. In k steps machine M can fill no more than k additional cells, and hence for $i = 0, \ldots, k$,

$$p_i < \beta^{l-2}, \qquad t_i < \beta^{l-2}; \tag{5.5.23}$$

in particular,

$$p' < \beta^{l-2}, \qquad t' < \beta^{l-2}. \tag{5.5.24}$$

Let us construct a pair of *superconfigurations* $\langle p_{\mathrm{L}}, t_{\mathrm{L}} \rangle$ and $\langle p_{\mathrm{R}}, t_{\mathrm{R}} \rangle$ by concatenating the configurations (5.5.18) (recall that in Section 3.3 we showed that concatenation of tuples with equal bases is a Diophantine operation):

$$\langle p_{\mathrm{L}}, \beta, kl \rangle = \langle p_0, \beta, l \rangle + \cdots + \langle p_{k-1}, \beta, l \rangle, \tag{5.5.25}$$

$$\langle t_{\mathrm{L}}, \beta, kl \rangle = \langle t_0, \beta, l \rangle + \cdots + \langle t_{k-1}, \beta, l \rangle, \tag{5.5.26}$$

$$\langle p_{\mathrm{R}}, \beta, kl \rangle = \langle p_1, \beta, l \rangle + \cdots + \langle p_k, \beta, l \rangle, \tag{5.5.27}$$

$$\langle t_{\mathrm{R}}, \beta, kl \rangle = \langle t_1, \beta, l \rangle + \cdots + \langle t_k, \beta, l \rangle. \tag{5.5.28}$$

The pairs $\langle p_{\mathrm{L}}, t_{\mathrm{L}} \rangle$ and $\langle p_{\mathrm{R}}, t_{\mathrm{R}} \rangle$ do not code configurations because the tuples with ciphers p_{L} and p_{R} contain k non-zero elements and the tuples with ciphers t_{L} and t_{R} correspond to tapes with k cells marked by the symbol "$\star$".

The superconfigurations correspond to a *supermachine*, whose tape is divided by the symbols "$\star$" into k parts on each of which a separate head operates according to the instructions of the original machine M. Instead of k steps of the machine M, we can deal with a single step of the supermachine, during which each of the heads carries out the corresponding instruction.

Thus, the superconfiguration $\langle p_{\mathrm{R}}, t_{\mathrm{R}} \rangle$ is uniquely determined by the superconfiguration $\langle p_{\mathrm{L}}, t_{\mathrm{L}} \rangle$. Moreover, the functions NextP and NextT, as defined by (5.5.16) and (5.5.8) without modification, already describe the operation of the superma- chine:

$$p_{\mathrm{R}} = \mathrm{NextP}(p_{\mathrm{L}}, t_{\mathrm{L}}), \qquad t_{\mathrm{R}} = \mathrm{NextT}(p_{\mathrm{L}}, t_{\mathrm{L}}). \tag{5.5.29}$$

Let us also consider the superconfiguration $\langle p_{\mathrm{M}}, t_{\mathrm{M}} \rangle$ defined by

$$\langle p_{\mathrm{M}}, \beta, (k-1)l \rangle = \langle p_1, \beta, l \rangle + \cdots + \langle p_{k-1}, \beta, l \rangle, \tag{5.5.30}$$

$$\langle t_{\mathrm{M}}, \beta, (k-1)l \rangle = \langle t_1, \beta, l \rangle + \cdots + \langle t_{k-1}, \beta, l \rangle. \tag{5.5.31}$$

Using this notation, equations (5.5.25)–(5.5.26) can be rewritten as

$$\langle p_{\mathrm{L}}, \beta, kl \rangle = \langle p, \beta, l \rangle + \langle p_{\mathrm{M}}, \beta, (k-1)l \rangle, \tag{5.5.32}$$

$$\langle t_{\mathrm{L}}, \beta, kl \rangle = \langle t, \beta, l \rangle + \langle t_{\mathrm{M}}, \beta, (k-1)l \rangle, \tag{5.5.33}$$

$$\langle p_{\mathrm{R}}, \beta, kl \rangle = \langle p_{\mathrm{M}}, \beta, (k-1)l \rangle + \langle p', \beta, l \rangle, \tag{5.5.34}$$

$$\langle t_{\mathrm{R}}, \beta, kl \rangle = \langle t_{\mathrm{M}}, \beta, (k-1)l \rangle + \langle t', \beta, l \rangle. \tag{5.5.35}$$

We have seen that for any configuration $\langle p, t \rangle$ and any positive k, if l satisfies (5.5.22), then there are numbers p_{L}, t_{L}, p_{M}, t_{M}, p_{R}, t_{R}, p', t' satisfying conditions (5.5.24), (5.5.29), and (5.5.32)–(5.5.35). We now show that these numbers are uniquely determined by these same conditions in terms of k, l, p, and t.

In fact, by (5.5.32) and (5.5.33), the first l elements of the tuples with codes $\langle p_{\mathrm{L}}, b, kl \rangle$ and $\langle t_{\mathrm{L}}, \beta, kl \rangle$ are uniquely determined. The functions NextP and NextT were defined by (5.5.16) and (5.5.8) in such a manner that the first m elements of the tuples with ciphers $\mathrm{NextP}(x, y)$ and $\mathrm{NextT}(x, y)$ are uniquely determined by the first $m+1$ elements of the tuples with ciphers x and y. Hence, by (5.5.29), the first $l-1$ elements of the tuples with codes $\langle p_{\mathrm{R}}, \beta, kl \rangle$ and $\langle t_{\mathrm{R}}, \beta, kl \rangle$ are also uniquely determined. Then it follows from (5.5.34) and (5.5.35) that the first $l-1$ elements of the tuples with codes $\langle p_{\mathrm{M}}, \beta, (k-1)l \rangle$ and $\langle t_{\mathrm{M}}, \beta, (k-1)l \rangle$ are also unique.

Returning to (5.5.32) and (5.5.33), we see that the first $2l-1$ elements of the tuples with codes $\langle p_{\mathrm{L}}, \beta, kl \rangle$ and $\langle t_{\mathrm{L}}, \beta, kl \rangle$ are uniquely determined. Hence, by (5.5.29), the first $2l-2$ elements of the tuples with codes $\langle p_{\mathrm{R}}, \beta, kl \rangle$ and $\langle t_{\mathrm{R}}, \beta, kl \rangle$ are unique, and, by (5.5.34) and (5.5.35), the first $2l-2$ elements of the tuples with codes $\langle p_{\mathrm{M}}, \beta, (k-1)l \rangle$ and $\langle t_{\mathrm{M}}, \beta, (k-1)l \rangle$ are unique as well.

Repeating this reasoning sufficiently often, we establish the uniqueness of all elements of the tuples with codes

$$\langle p_{\mathrm{L}}, \beta, kl \rangle, \langle t_{\mathrm{L}}, \beta, kl \rangle, \langle p_{\mathrm{M}}, \beta, (k-1)l \rangle, \langle t_{\mathrm{M}}, \beta, (k-1)l \rangle$$

and all, except possibly the last, elements of the tuples with codes

$$\langle p_{\mathrm{R}}, \beta, kl \rangle, \langle t_{\mathrm{R}}, \beta, kl \rangle, \langle p', \beta, l \rangle, \langle t', \beta, l \rangle.$$

In addition, the inequality (5.5.24) implies that the last elements of the tuples with codes $\langle p', \beta, l \rangle$ and $\langle t', \beta, l \rangle$ are equal to 0, and hence, by (5.5.34) and (5.5.35), so are the last elements of the tuples with codes $\langle p_{\mathrm{R}}, \beta, kl \rangle$ and $\langle t_{\mathrm{R}}, \beta, kl \rangle$.

Thus, we have shown that *for any configuration* $\langle p, t \rangle$, *any positive* k, *and for every* l *satisfying* (5.5.22), *the system of Diophantine conditions* (5.5.24), (5.5.29), (5.5.32)–(5.5.35) *has exactly one solution in the variables* p_L, t_L, p_M, t_M, p_R, t_R, p', t', *and for this unique solution, we have* $p' = \mathrm{AfterP}(k, p, t)$ *and* $t' = \mathrm{AfterT}(k, p, t)$. This immediately implies that the functions AfterP and AfterT are Diophantine.

To construct the Diophantine equation (5.5.4), we need to take one additional small step. Let $\omega_1, \ldots, \omega_z$ be the subscripts of the final states of machine M. Now, the condition

$$\exists kr \left[\mathrm{Elem}\big(\mathrm{AfterT}(k,p,t), \beta, r\big) = \omega_1 \vee \cdots \vee \mathrm{Elem}\big(\mathrm{AfterT}(k,p,t), \beta, r\big) = \omega_z\right] \tag{5.5.36}$$

holds for the code $\langle p, t \rangle$ of a configuration if and only if the machine, beginning in that configuration, will eventually halt (with k being equal to the number of steps carried out by the machine). Condition (5.5.36) is Diophantine and hence can be transformed into the required equation (5.5.4).

However, equation (5.5.4) is not quite what we need in order to prove that the set $\mathfrak{M}$, semidecidable by machine M, is Diophantine, because the parameters of (5.5.4) are p and t from the code of a configuration, rather than $a_1, \ldots, a_n$ as required in a Diophantine representation of a set. Nevertheless, it is easy to see that in the initial configuration,

$$p = 1 \tag{5.5.37}$$

(the machine is in state q_1 with its head scanning the leftmost cell) and

$$\langle t, \beta, a \rangle = \langle \kappa, \beta, 1 \rangle + \langle \mu, \beta, 1 \rangle + \langle \mathrm{Repeat}(\nu, \beta, a_1), \beta, a_1 \rangle + \cdots + \langle \mu, \beta, 1 \rangle + \langle \mathrm{Repeat}(\nu, \beta, a_n), \beta, a_n \rangle, \tag{5.5.38}$$

where $a = a_1 + \cdots + a_n + n + 1$ is the number of occupied cells and κ, μ, and ν are the subscripts of the symbols "$\star$", "0", and "1", respectively in (5.5.1) (i.e., $\alpha_\kappa = \star$, $\alpha_\mu = 0$, $\alpha_\nu = 1$). Combining (5.5.4), (5.5.37), and (5.5.38) and regarding $a_1, \ldots, a_n$ as parameters, we obtain the required Diophantine representation of the set $\mathfrak{M}$.

5.6 Hilbert's Tenth Problem is undecidable by Turing machines

In Section 5.5 we introduced the notion of Turing semidecidable set, and we established the equivalence of that notion with the notion of Diophantine set. Now, as soon as we introduce one other notion, we shall be in a position to give a more precise formulation of Hilbert's Tenth Problem. Namely, we shall say that a set $\mathfrak{M}$ of n-tuples of natural numbers is *Turing decidable* if there is a Turing machine M that, beginning in state q_1, with the head scanning the leftmost cell of a tape containing the canonical representation of the tuple $\langle a_1, \dots, a_n \rangle$, will eventually halt in state q_2 if $\langle a_1, \dots, a_n \rangle \in \mathfrak{M}$ and in state q_3 otherwise.

It is natural to ask what relationship there is between Turing decidability and semidecidability. It is easy to see that *if a set $\mathfrak{M}$ is Turing decidable, then it is Turing semidecidable.* In fact, if M is a Turing machine yielding the decidability of $\mathfrak{M}$, then the machine

$$\textbf{while } M \textbf{ do } \text{STOP } \textbf{od}; \text{NEVERSTOP} \tag{5.6.1}$$

semidecides the set $\mathfrak{M}$.

It is also easy to see that *if a set is decidable, then its complement is also semidecidable.* For this, it suffices to consider the machine

$$\textbf{while } M \textbf{ do } \text{NEVERSTOP } \textbf{od} \tag{5.6.2}$$

instead of (5.6.1).

It is not so evident that the converse is valid too: *if both a set $\mathfrak{M}$ and its complement are Turing semidecidable, then $\mathfrak{M}$ is Turing decidable.* A straightforward approach to a proof of this result might go as follows. Let M_2 and M_3 be Turing machines that semidecide the set $\mathfrak{M}$ and its complement, respectively. One can construct a third machine M that simulates the operations of the two machines M_2 and M_3 on its single tape, performing in turn the successive steps of each machine. (Naturally, simulating one step of M_2 or M_3 requires machine M to perform many steps.) Eventually, one of the two machines, M_2 or M_3, will halt, and then machine M will enter the final state q_2 or q_3, respectively.

The technical details involved in actually constructing a machine that simultaneously simulates two given machines are quite complicated. However, taking advantage of the equivalence, established above, of the class of Turing semidecidable sets with the class of Diophantine sets, we can give a shorter proof. Instead of simulating Turing machines, we construct a pair of Diophantine equations

$$D_2(a_1, \ldots, a_n, x_1, \ldots, x_m) = 0 \tag{5.6.3}$$

and

$$D_3(a_1, \ldots, a_n, x_1, \ldots, x_m) = 0, \tag{5.6.4}$$

respectively defining the set $\mathfrak{M}$ and its complement (without loss of generality, we assume that both equations have the same number of unknowns). The two equations can be combined into a single equation

$$(D_2^2(a_1, \ldots, a_n, x_1, \ldots, x_m) + (1 - y_m)^2) \cdot (D_3^2(a_1, \ldots, a_n, x_1, \ldots, x_m) + y_m) = 0. \tag{5.6.5}$$

Clearly, this equation has a solution for any value of the parameters, and moreover, $\langle a_1, \ldots, a_n \rangle \in \mathfrak{M}$ if and only if

$$y_m = 1 \tag{5.6.6}$$

in such a solution. Let M be machine (5.4.10) constructed with equation (5.6.5) substituted for equation (5.4.1). When M halts, the tape will contain the representation of a tuple of the form (5.4.9). A machine to decide $\mathfrak{M}$ can then be defined as $M;\text{EQUAL}(l, l+1)$, where $l = n + 1 + 2m$; this machine determines whether equation (5.6.6) holds because the constant 1 and the value of y_m are the lth and $l + 1$st elements, respectively, in the tuple (5.4.9).

Now we can reformulate Hilbert's Tenth Problem in the following, more explicit way: *is the set of codes of all solvable Diophantine equations (without parameters) Turing decidable*? This set was considered in Section 4.5, where it was denoted by $\mathfrak{H}_0$. There we established that $\overline{\mathfrak{H}_0}$, the complement of $\mathfrak{H}_0$, is not Diophantine. This implies that $\mathfrak{H}_0$ is not Turing decidable. In other words, *it is impossible to construct a Turing machine that, beginning with the representation of a number k on the tape, will halt after a finite number of steps in state* q_2 *or* q_3*, depending on whether the equation with code k is or is not solvable.*

The set $\mathfrak{H}_0$ is itself Diophantine, since it is defined by equation (4.6.7). This fact shows that Turing machines are incapable of deciding whether or not the equations belonging *to one particular family of Diophantine equations* have solutions, to say nothing of *arbitrary Diophantine equations.*

5.7 Church's Thesis

In the previous sections of this chapter we have obtained two, no doubt remarkable, results. Namely, we have established that

$$\text{\textit{the class of Diophantine sets}} \\ \text{is identical to} \\ \text{\textit{the class of Turing semidecidable sets}} \tag{5.7.1}$$

and

$$\text{\textit{Hilbert's Tenth Problem is Turing undecidable.}} \tag{5.7.2}$$

However, these two rather technical results raise a number of new questions. While the definition of *Diophantine set* is quite natural, that of *Turing semidecidable set* is burdened by numerous technical details, many of which seem arbitrary. For example, we could use binary rather than unary notation to represent numbers. The tape could be infinite in both directions instead of in only one direction. Instead of a single head, there might be several, each executing its own set of instructions while sharing information about their respective scanned cells. Moreover, there might be several tapes. In fact, the memory need not even be linear; it could, for example, take the form of a plane divided into square cells. For each such modification of the notion of Turing machine, one can introduce a corresponding concept of semidecidability and pose the question of how that concept is related to Diophantine sets.

The reformulation of Hilbert's Tenth Problem used in Section 5.6 could be criticized as well, since it is based on a very special method for coding Diophantine equations. It would be more natural to write on the tape the number of unknowns, the degree, and the coefficients of an equation in unary or some positional notation. Hilbert did not impose any restrictions on the desired method for solving the Tenth Problem. Thus, if for some appropriate notation for polynomials, someone had succeeded in constructing a decision machine, that would certainly have provided a positive solution of Hilbert's Tenth Problem. So to what extent can the Turing undecidability established in Section 5.6 be considered as constituting a negative solution?

For every modification of the notion of Turing machine (or any other abstract computing device) and for every method of representing initial data on the tape, one could try to obtain results like (5.7.1) and (5.7.2). This could be done directly or indirectly. For the former, we would need to introduce suitable codings and to prove that the corresponding relations are Diophantine. For the latter, we can forget about Diophantine equations for the time being and prove that the class of semidecidable sets remains the same in spite of modifications in the definition (see, e.g., Exercises 5.6 and 5.7a). Such studies of relations between (semi)decidability on different abstract computing devices were in fact carried out quite apart from work on Hilbert's Tenth Problem, and long before its undecidability was established in any sense. These studies showed that the notions of semidecidable set and decidable set, which we introduced using Turing machines, are quite insensitive to the choice of a particular computing device and method of representing initial data. The very same classes arose for any "reasonable" choice. By "reasonable" we mean that the elementary steps of a computation should in fact be elementary, i.e., easily performable (for example, one cannot recognize in a single step whether a particular Diophantine equation has a solution).

Long before the appearance of the first mathematically rigorous notions of abstract computing devices such as Turing machines, there was the intuitive notion of an algorithm as a guaranteed method for the mechanical solution of problems of a definite kind. As a classical example, we can mention Euclid's algorithm for finding the greatest common divisor of two positive integers. Whenever an algorithm was proposed, it was clear that in fact it was a universal method for solving problems of the appropriate kind. However, such intuitive notions do not suffice for a proof that there is no algorithm for problems of the kind being considered.

The initial data for an algorithm (in the intuitive sense) are selected from some countable set, and, essentially without loss of generality, we are only going to consider situations in which the initial data consist of natural numbers or tuples of natural numbers of some fixed length. (The techniques of Chapter 4 indicate how one could code more complicated structures.)

The result of carrying out an algorithm is also an object of the appropriate kind. We could have chosen to consider only algorithms whose outputs are natural numbers, but it is more in the spirit of the subject of this book to consider instead algorithms with the two outputs "YES" and "NO." Correspondingly, along with the intuitive concept of algorithm, two related concepts arise, namely, the intuitive concepts of decidable and semidecidable set. A set $\mathfrak{M}$ of n-tuples is decidable (in the intuitive sense) if there exists an algorithm (also in the intuitive sense) that halts on every n-tuple of natural numbers and reports "YES" or "NO" depending on whether the n-tuple belongs to the set or not. Similarly, for $\mathfrak{M}$ to be semidecidable we would need an algorithm that would report "YES" for every n-tuple in $\mathfrak{M}$ and that would either report "NO" or fail to halt if the tuple doesn't belong to $\mathfrak{M}$. Computability theory in its entirety could be expounded in terms of decidable and semidecidable sets, just as we can eliminate the notion of functions from mathematics and deal only with their graphs.

Between intuitive decidability and semidecidability there is the same relation as between Turing decidability and semidecidability: a set is decidable if and only if both it and its complement are semidecidable.

How is formal Turing (semi)decidability related to (semi)decidability in the intuitive sense? One relation is evident: Turing (semi)decidable sets are recognized by most mathematicians as (semi)decidable in the intuitive sense. The converse is known as

CHURCH'S THESIS: *Every set of n-tuples that is (semi)decidable in the intuitive sense is also Turing (semi)decidable.*

Here we have encountered something rarely found in mathematics, a *thesis*. What is it? It is not a theorem because it has no proof. It is not a conjecture because it cannot have a proof. It is not even an axiom that we are free to accept or reject. All of this is due to the fact that Church's Thesis is not a precise mathematical statement, because it relates the rigorous notion of Turing (semi)decidability with the non-rigorous notion of (semi)decidability in the intuitive sense.

What are the arguments in favor of Church's Thesis? They may be very different in appearance but all of them are essentially the same: *so far no one has found an example of a set that would be recognized by most mathematicians as (semi)decidable in the intuitive sense and for which we cannot construct an appropriate Turing machine.* Of course, this assertion should not be understood literally, i.e., as though for every set recognized to be (semi)decidable someone had actually constructed a corresponding machine. In fact, there are numerous tools for indirectly proving the existence of such a machine.

What is Church's Thesis for? On the one hand, it can serve as a guiding star: as soon as we have established (semi)decidability in the intuitive sense, our chance of finding the corresponding Turing machine should be considered very high. In fact, professional mathematicians usually content themselves with establishing intuitive (semi)decidability and do not condescend to formal proofs.

On the other hand, Church's Thesis plays a role in mathematics similar to that played elsewhere by the law of conservation of energy. Namely, as long as no exception to the law is found, it is unreasonable to set out to construct a perpetual motion machine. Similarly, once the Turing undecidability of a set has been proved, one need not spend one's time seeking a universal method for recognizing the elements of that set. In particular, according to Church's Thesis, the quite special result (5.7.2) gives us the moral right to cease the hunt (so far carried out in vain) for a "process" of the sort Hilbert asked for in his Tenth Problem.

In Chapter 7 we shall discuss the question of whether (5.7.2) constitutes a complete solution of Hilbert's Tenth Problem from another point of view.

Exercises

1. Construct a Turing machine DOUBLE that will transform tuple (5.3.1) into the tuple $\langle a_1, \dots, a_{n-1}, 2a_n \rangle$ for any n.

2. Construct a Turing machine ERASE that will transform tuple (5.3.1) into the tuple $\langle a_2, \dots, a_n \rangle$.

3. Let the *binary representation of a number* on tape consist of consecutive cells containing the symbols "0" and "1" (with the most significant digits to the left) followed by a cell containing the symbol ",", and let the *binary representation of a tuple* consist of the binary representations of its elements in order without gaps. Construct, for each n, a Turing machine BIN(n) that will transform the binary representation of an n-tuple into its unary representation.

4. Using the binary representation of tuples from Exercise 5.3, one can introduce the notion of a *binary Turing semidecidable set* in a natural manner. The existence of the machines BIN shows that every Diophantine set is binary Turing semidecidable. Show that the converse is true as well.

5. Let us say that a natural number valued function F defined on a set of n-tuples of natural numbers is *Turing computable* if there is a Turing machine M that, beginning with the canonical representation of the tuple $\langle a_1, \dots, a_n \rangle$ on its tape, will eventually halt if and only if $F(a_1, \dots, a_n)$ is defined, and for which, moreover, the tape will contain a representation of the singletong $\langle F(a_1, \dots, a_n) \rangle$ when it halts. Show that a function is Turing computable if and only if it is Diophantine.

6. Show that every set (semi)decidable by a Turing machine with a tape that is infinite in both directions is also (semi)decidable by a Turing machine with a tape possessing an end.

7. After the introduction of Turing machines, many other kinds of abstract computing devices were proposed. Certain of these can be used instead of Turing machines for establishing the relationship between "Diophantine" and "semidecidable." As an example, we can mention *register machines.* The memory of such a machine consists of finitely many *registers,* each of which can contain an arbitrarily large natural number. The behavior of a machine is specified by a program consisting of finitely many labeled *instructions.* The instructions may be of four types:

I. Lj: R$i \leftarrow$ R$i + 1$;
II. Lj: R$i \leftarrow$ R$i - 1$;
III. Lj: **if** R$i = 0$ **then goto** Lk;
IV. Lj: **stop**.

An instruction of type I increases the contents of a register by 1; an instruction of type II decreases it by 1. An instruction of type III is a conditional branch breaking the natural order in which instructions are executed. Finally, an instruction of type IV stops the machine. It is assumed that programs are written in such a way that instructions of type II will execute only when the corresponding register does not contain 0. The notion of a *set semidecidable by a register machine* is introduced in a natural way: the elements of an n-tuple are placed into n input registers (while the other registers contain 0), the machine begins by executing the first instruction, and the n-tuple belongs to the set if and only if the machine eventually halts.

(a) Prove that the notions of Turing semidecidability and semidecidability on register machines are equivalent.
(b) Give a direct proof of the equivalence between "Diophantine" and "semidecidable by a register machine" by simulating register machines by Diophantine equations.

9. Show that every Diophantine set can be represented in the form

$$a \in \mathfrak{D} \iff \exists x_1 \dots x_m \left[\sum_{k=1}^{l} \lambda(k) x_1^{P_{k1}(a)} \dots x_m^{P_{km}(a)} = 0 \right],$$

where P_{kj} are polynomials with natural number coefficients and $\lambda(k) = \pm 1$.

Commentary

Alan Turing introduced the abstract computing devices that are now named for him in his classical paper [1936]. A very similar notion was also introduced by Emil L. Post [1936]. Since that time numerous modifications of the Turing-Post machine have been proposed. The version used in Section 5.1 was chosen as being particularly suitable for simulation by Diophantine equations.

Various authors have proposed other approaches for making the general notion of algorithm precise. (An exposition of the history of computability theory (in Russia the subject is often referred to as *the theory of algorithms*) can be found in Uspenskiĭ and Semënov [1987].) All of these approaches led to equivalent notions of decidable and semidecidable sets (the latter are more often called "recursively enumerable"). Alonzo Church [1936] was the first to realize that a single and, at first sight, very special definition can be adequate for the fundamental notion of computability. Church's Thesis has many equivalent formulations depending on the choice of a particular kind of abstract computing device, and the formulation from Section 5.7 is sometimes called *Turing's Thesis.* Kolmogorov and Uspenskiĭ [1958] attempted to give the most general definition of an abstract computing device satisfying the requirement that every step should be elementary, and proved its equivalence to more traditional models of computing devices, in particular to Turing machines.

The first papers aimed at proving the algorithmic unsolvability of Hilbert's Tenth Problem appeared in the early 1950's (see the Commentary to Chapter 1). Even at that time, there was no difficulty in proving that all Diophantine sets are semidecidable (for any standard definition of the latter notion). At this same time Martin Davis [1953] set forth the daring hypothesis that the converse is also true, i.e., that *every semidecidable set is Diophantine,* and thus that the number-theoretic notion of Diophantine set coincides with the notion of semidecidable set from computability theory.

Davis [1950, 1953] took the first step towards proving his hypothesis; namely, he showed that every semidecidable set of natural numbers can be represented in the form

$$a \in \mathfrak{M} \iff \exists z \forall y \le z \, \exists x_1 \ldots x_m \, [D(a, x_1, \ldots, x_m, y, z) = 0]. \tag{1}$$

Such representations are said to be in *Davis normal form.*

Examples of sets that are semidecidable but not decidable have been known since the 1930's, and hence Davis's hypothesis implied a negative solution of Hilbert's Tenth Problem. However, efforts to prove the hypothesis encountered difficulties.

In this situation, it was natural to seek other approaches that, without proving Davis's hypothesis, would nevertheless provide a negative solution of Hilbert's Tenth Problem.

One possible way to weaken Davis's hypothesis is as follows: In the definition of a family of equations given in Section 1.4, it is required that the function D in (1.4.1) should be a polynomial in *all* the variables $a_1, \ldots, a_n, x_1, \ldots, x_m$. Instead of this, it would be sufficient to require polynomial dependence on $x_1, \ldots, x_m$ only. It would suffice that the dependence on $a_1, \ldots, a_n$ be via any algorithm that would provide, for any given values of $a_1, \ldots, a_n$, a Diophantine equation in $x_1, \ldots, x_m$. It is clear that even under such a weakened restriction on D, a positive solution of Hilbert's Tenth Problem would have implied the decidability of the set represented by (1.4.2).

In particular, one could consider the class of exponential Diophantine equations in which unknowns do not occur in the exponents. Such an approach was proposed by A. I. Mal'tsev [1968], but for reasons that are not entirely clear, although he allowed parameters to occur in the exponents, he did not permit parameters to occur polynomially like the unknowns. Thus, Davis's hypothesis does not formally give a solution to the problem posed by Mal'zev; however, a solution can be easily deduced from it (see Exercise 5.9).

Another weakening of Davis's hypothesis was proposed by S. Yu. Maslov [1967].

Davis's hypothesis was finally proved in 1970, but the chronological order in which the proof was obtained was different from the logical order followed in Chapters 1–5. One of the most crucial steps in the historical development consisted in obtaining an exponential Diophantine representation for an arbitrary semidecidable set. This result was first obtained by Davis and Putnam in a conditional form. Their result was announced in an abstract [1959a] and presented with full details in a report [1959b] that is not readily available.

In their proof Davis and Putnam used the still-unproved (1992) conjecture that there exist arbitrarily long arithmetical progressions consisting entirely of prime numbers. It was Julia Robinson who managed to eliminate the conjecture. The new result was announced in her short communication [1960] and a lecture [1962] and then published in the now-classic paper by Davis, Putnam, and Robinson [1961]. In that paper, an exponential Diophantine representation was obtained by eliminating the bounded universal quantifier from the Davis normal form (1). This technique will be presented in Chapter 6.

After it was established that every semidecidable set has an exponential Diophantine representation, it was sufficient to show that exponentiation was Diophantine in order to prove Davis's hypothesis. The history of that stage was presented in the Commentary to Chapter 2.

The proof that exponentiation was Diophantine was not only the final link in the proof of Davis's hypothesis, it also opened new routes for the construction of Diophantine representations of semidecidable sets. In particular, the Davis representation (1) ceased being a necessary step. Earlier it had been necessary because the method used for elimination of the bounded universal quantifier preceding the Diophantine equation in (1) led to an exponential Diophantine equation, and thus the process could not be iterated. Now that we can transform exponential Diophantine equations into Diophantine equations, we can begin with an arithmetic representation of a semidecidable set containing arbitrarily many universal quantifiers. A technique for obtaining arithmetical representations had been found long before by Kurt Gödel [1931].

On the other hand, the fact that exponentiation was Diophantine also made it possible to construct Diophantine representations without any use of bounded universal quantifiers, as was in fact done in Chapters 1–5. An approach of this kind was implemented for the first time by Matiyasevich [1976]. This construction was based on Turing machines, and the main difficulties were connected with the necessity of coding strings of symbols and sequences of their transformations by numbers. In Jones and Matiyasevich [1983, 1984, 1991] and Matiyasevich [1984], another construction is presented in which the initial representation of a semidecidable set is based on register machines (see Exercise 5.7). This model of abstract computing devices was proposed in Lambek [1961], Melzak [1961], Minsky [1961, 1967], and Shepherdson and Sturgis [1963]. Register machines are particularly suitable for constructing Diophantine representations. Like Turing machines, they have very primitive instructions and, in addition, they deal directly with numbers.

The method of Diophantine simulation of Turing machines described in Section 5.5 is different from the method given in Matiyasevich [1976] and is presented here for the first time. For yet another proof via Turing machines, see van Emde Boas [1983].

We conclude with a quotation from Hilbert's lecture [1900] that reveals his attitude to "negative" results:

> Occasionally it happens that we seek the solution under insufficient hypotheses or in an incorrect sense, and for this reason do not succeed. The problem then arises: to show the impossibility of the solution under the given hypotheses, or in the sense contemplated. Such proofs of im-

possibility were effected by the ancients, for instance when they showed that the ratio of the hypotenuse to the side of an isosceles triangle is irrational. In later mathematics, the question as to the impossibility of certain solutions plays a preëminent part, and we perceive in this way that old and difficult problems, such as the proof of the axiom of parallels, the squaring of circle, or the solution of equations of the fifth degree by radicals have finally found fully satisfactory and rigorous solutions, although in another sense than that originally intended. It is probably this important fact along with other philosophical reasons that gives rise to conviction (which every mathematician shares, but which no one has as yet supported by a proof) that every definite mathematical problem must necessary be susceptible of an exact settlement, either in the form of an actual answer to the question asked, or by the proof of the impossibility of its solution and therewith the necessary failure of all attempts.

6 Bounded Universal Quantifiers

In Sections 6.1–6.3, we explain three different methods for transforming formulas containing bounded universal quantifiers into equivalent formulas containing only existential quantifiers. In Sections 6.4–6.6, various examples serve to demonstrate the power of this technique for constructing Diophantine representations.

6.1 First construction: Turing machines

We are to show that if $\mathcal{R}(a_1, \ldots, a_n)$ is a Diophantine relation, then so is the relation defined by the formula

$$\forall y < a_n \mathcal{R}(a_1, \ldots, a_{n-1}, y). \tag{6.1.1}$$

First, using Church's Thesis, we present heuristic evidence that this is indeed the case. In order to do so, we suppose that $\mathcal{R}$ is semidecidable and show how to produce an algorithm (in the intuitive sense) for testing the truth of (6.1.1) for given values of a_1, ..., a_n. By definition, formula (6.1.1) is vacuously true if $a_n = 0$; for $a_n > 0$ it is equivalent to the statement that each of the following assertions is true:

$$\begin{gathered} \mathcal{R}(a_1, \ldots, a_{n-1}, 0), \\ \vdots \\ \mathcal{R}(a_1, \ldots, a_{n-1}, a_n - 1). \end{gathered} \tag{6.1.2}$$

By our assumption that $\mathcal{R}$ is semidecidable, we can test the validity of each of these assertions in a finite number of steps and thus are able to test the validity of (6.1.1).

This argument is not a formal proof, and it provides no method for transforming (6.1.1) into an equivalent Diophantine equation, even if we possess a Diophantine representation for $\mathcal{R}$. In order to proceed formally, we need to describe the corresponding Turing machines.

Thus, let $\mathcal{R}$ be defined by the Diophantine equation

$$D(a_1, \ldots, a_n, x_1, \ldots, x_{m+1}) = 0. \tag{6.1.3}$$

Let us regard this equation as being equation (5.4.1), and using the method of Section 5.4, let us construct the machine M that semidecides $\mathcal{R}$. When M halts, the tape will contain the tuple (5.4.9). Let us define machine M' by

$$M' = M; M_2; \text{DELETE} \tag{6.1.4}$$

where M_2 is defined as in Section 5.4. Machine M' also semidecides $\mathcal{R}$, but when it halts, the tape contains the initial tuple $\langle a_1, \ldots, a_n \rangle$.

Now it is easy to construct a machine M'' that will semidecide (6.1.1):

$$M'' = \textbf{while } \text{DEC } \textbf{do } M' \textbf{ od} \tag{6.1.5}$$

Clearly, this machine works as follows. If $a_n = 0$, then machine DEC will halt in state q_3, and hence machine M'' will halt in state q_2. If $a_n > 0$, then machine M' will start by testing the assertion $\mathcal{R}(a_1, \ldots, a_{n-1}, a_n - 1)$. If the result is positive, then the machine DEC will determine whether $a_n - 1 = 0$, and if not, machine M' will begin testing the assertion $\mathcal{R}(a_1, \ldots, a_{n-1}, a_n - 2)$ and so on. If each of the assertions (6.1.2) is true, then machine M'' will eventually halt in state q_2.

To obtain a Diophantine equation

$$E(a_1, \ldots, a_n, x_1, \ldots, x_w) = 0 \tag{6.1.6}$$

equivalent to formula (6.1.1), it suffices to simulate machine M'' using the method of Section 5.5.

The method just described for transforming equation (6.1.3) into equation (6.1.6) is quite constructive, but rather roundabout. In the following sections we introduce more straightforward number-theoretic methods for obtaining this result.

6.2 Second construction: Gödel coding

To simplify the notation, we give the proof for the case when $n = 2$ in (6.1.1) and, correspondingly, in (6.1.3). The proof for the general case is exactly the same; another approach would be to go from the case $n = 2$ to the general case by using Cantor numbering (cf. Section 4.4).

So, we have to show that the set of pairs defined by the formula

$$\forall y < b \, \exists x_1 \ldots x_m \, [D(a, y, x_1, \ldots, x_m) = 0] \tag{6.2.1}$$

is Diophantine; here D is, as usual, an arbitrary polynomial with integral coefficients. In other words, we have to find another polynomial E such that formula (6.2.1) is equivalent to the formula

$$\exists z_0 \ldots z_k \, [E(a, b, z_0, \ldots, z_k) = 0]. \tag{6.2.2}$$

The validity of (6.2.2) for particular values of a and b is supposed to imply the validity of (6.2.1), i.e., the existence of b tuples

$$\begin{gathered}\langle x_{1,0}, \ldots, x_{m,0} \rangle, \\ \vdots \\ \langle x_{1,b-1}, \ldots, x_{m,b-1} \rangle \end{gathered} \tag{6.2.3}$$

such that for any y between 0 and $b-1$,

$$D(a, y, x_{1,y}, \ldots, x_{m,y}) = 0. \tag{6.2.4}$$

It is natural to expect that information concerning the numbers (6.2.3) should be contained in some form in the numbers $z_0, \ldots, z_k$ whose existence is asserted in (6.2.2).

Instead of (6.2.3), *an indefinite number of tuples of fixed length*, we consider *a fixed number of tuples of indefinite length*

$$\begin{gathered}\langle x_{1,0}, \ldots, x_{1,b-1} \rangle, \\ \vdots \\ \langle x_{m,0}, \ldots, x_{m,b-1} \rangle. \end{gathered} \tag{6.2.5}$$

To deal with (6.2.5), we shall use a coding closely related to the Gödel coding considered in Section 3.2. Namely, we choose pairwise coprime numbers $q_0, \ldots, q_{b-1}$ such that for $i = 1, \ldots, m$ and $y = 0, \ldots, b-1$,

$$x_{i,y} = \operatorname{rem}(z_i, q_y). \tag{6.2.6}$$

The number z_0 will satisfy the condition

$$y = \operatorname{rem}(z_0, q_y) \tag{6.2.7}$$

for $y = 0, \ldots, b-1$. Codings based on the Chinese Remainder Theorem have the following advantage: they make it possible to calculate the value of a polynomial directly from the codes without any preliminary decoding. Namely,

$$D(a, y, x_{1,y}, \ldots, x_{m,y}) \equiv D(a, z_0, z_1, \ldots, z_m) \pmod{q_y}, \tag{6.2.8}$$

and hence if q_y is so large that, in addition to (6.2.6), we have also the inequality

$$|D(a, y, x_{1,y}, \ldots, x_{m,y})| < q_y, \tag{6.2.9}$$

then the validity of (6.2.4) for some y is equivalent to the validity of the congruence

$$D(a, z_0, z_1, \ldots, z_m) \equiv 0 \pmod{q_y}. \tag{6.2.10}$$

In (6.2.10) only the modulus depends on y, and hence the validity of (6.2.4) for all values of y from 0 to $b-1$ is equivalent to the validity of the single congruence

$$D(a, z_0, z_1, \ldots, z_m) \equiv 0 \pmod{q_0 \cdots q_{b-1}}. \tag{6.2.11}$$

So far we haven't eliminated the bounded universal quantifier entirely; for example, in (6.2.11) it is hidden in the three dots. To eliminate the bounded universal quantifier from (6.2.11), we take advantage of our freedom in choosing $q_0, \ldots, q_{b-1}$ and select them in such a manner that the product $q_0 \cdots q_{b-1}$ will be a Diophantine function of b. Namely, in contrast to Gödel's choice (3.2.7) of the moduli, for $y = 0, \ldots, b-1$, we set

$$q_y = \frac{q+1}{y+1} - 1, \tag{6.2.12}$$

where q is large and $q+1$ is a multiple of $(b!)^2$. Then

$$q_0 \cdots q_{b-1} = \prod_{y=0}^{b-1} \frac{q-y}{y+1} = \binom{q}{b}, \tag{6.2.13}$$

and in Section 3.4 we proved that the binomial coefficients are Diophantine. Now we can rewrite (6.2.10) in the form of a Diophantine condition

$$D(a, z_0, z_1, \ldots, z_m) \equiv 0 \pmod{\binom{q}{b}}. \tag{6.2.14}$$

We still haven't eliminated the universal quantifier entirely, because we need to ensure that for $y = 0, \ldots, b-1$, conditions (6.2.7) and (6.2.9) are satisfied, where $x_{i,y}$ is defined by (6.2.6). Taking advantage of our fortunate choice of q_y, it is easily verified that (6.2.7) is satisfied by taking

$$z_0 = q. \tag{6.2.15}$$

Strictly speaking, we shall not quite ensure that (6.2.9) is satisfied. Instead we will use substitutes for (6.2.9) and (6.2.10), namely, the inequality

$$|D(a, y, x_{1,y}, \ldots, x_{m,y})| < p_y \tag{6.2.16}$$

and the congruence

$$D(a, y, x_{1,y}, \ldots, x_{m,y}) \equiv 0 \pmod{p_y}, \tag{6.2.17}$$

where

$$p_y \mid q_y, \tag{6.2.18}$$

$$x_{i,y} = \text{rem}(z_i, p_y). \tag{6.2.19}$$

Clearly, this is sufficient for the validity of (6.2.4).

To obtain (6.2.16), we introduce an auxiliary variable w and demand that for $i = 1, \ldots, m,\ y = 0, \ldots, b-1$

$$x_{i,y} < w. \tag{6.2.20}$$

Clearly, for the given values of a and b, such a bound w can be found. Now, it is evident that

$$|D(a, y, x_{1,y}, \ldots, x_{m,y})| \leq B(a, b, w), \tag{6.2.21}$$

where the polynomial $B(a, b, w)$ is obtained from $D(a, y, x_1, \ldots, x_m)$ by changing the signs of all of its negative coefficients and systematically replacing y by b and $x_1, \ldots, x_m$ by w. In order to obtain (6.2.16), we try to arrange matters so that

$$p_y > B(a, b, w) \tag{6.2.22}$$

and the numbers $x_{i,y}$ defined by (6.2.19) satisfy (6.2.20).

It follows from (6.2.6) and (6.2.20) that

$$q_y \,\Bigg|\, \prod_{k=0}^{w-1} (z_i - k). \tag{6.2.23}$$

We choose q so that

$$b!(b + w + B(a, b, w))! \mid q + 1; \tag{6.2.24}$$

then q_y is coprime with $w!$ and (6.2.23) will imply that

$$q_y \,\Bigg|\, \binom{z_i}{w}. \tag{6.2.25}$$

Finally, using (6.2.13), the system of bm conditions (6.2.25) can be compressed into the m Diophantine conditions

$$\binom{q}{b} \Bigm| \binom{z_1}{w},$$
$$\vdots \tag{6.2.26}$$
$$\binom{q}{b} \Bigm| \binom{z_m}{w}.$$

Thus, we have obtained a system of $m+3$ Diophantine conditions (6.2.14), (6.2.15), (6.2.24), and (6.2.26) that is solvable in the unknowns q, w, $z_0, \ldots, z_m$ provided that (6.2.1) holds. Now we need to verify that the converse is also true, i.e., that if the system consisting of (6.2.14), (6.2.15), (6.2.24), and (6.2.26) is solvable, then (6.2.1) holds.

Let y be any number less than b. Choose q_y according to (6.2.12). It follows from (6.2.26) that (6.2.25) is valid, and hence so is (6.2.23). Unfortunately, (6.2.23) does not imply (6.2.20) when $x_{i,y}$ are defined by (6.2.6) because q_y is, in general, composite. For this reason, we use p_y, an arbitrarily chosen *prime* factor of q_y, instead of q_y itself. Of course, (6.2.18) is then satisfied. It follows from (6.2.23) that

$$p_y \Bigm| \prod_{x=0}^{w-1} (z_i - x). \tag{6.2.27}$$

The fact that p_y is prime implies the existence of numbers $x_{1,y}, \ldots, x_{m,y}$ satisfying (6.2.20) such that

$$p_y \mid z_i - x_{i,y}. \tag{6.2.28}$$

From (6.2.15), (6.2.12), and (6.2.17), we see that

$$p_y \mid z_0 - y, \tag{6.2.29}$$

which together with (6.2.14) implies (6.2.10), and hence also that (6.2.17) is true. It is easy to see that by (6.2.24), all the prime factors of q_y are greater than $B(a, b, w)$; thus, (6.2.22) is true, and hence so is (6.2.16). As was already mentioned, (6.2.16) and (6.2.17) imply (6.2.4); since y was arbitrary, we see that (6.2.1) holds.

To obtain the desired Diophantine equation (6.2.2), it remains to eliminate the binomial coefficients and factorials from (6.2.14), (6.2.15), (6.2.24), and (6.2.26) (using tools from Section 3.4) and to combine the resulting Diophantine equations.

6.3 Third construction: summation

Once again we are tackling the same problem we dealt with in Sections 6.1 and 6.2; i.e., we are to show that the set of pairs defined by the formula

$$\forall y < b\, \exists x_1 \ldots x_m\, [D(a, y, x_1, \ldots, x_m) = 0] \tag{6.3.1}$$

is Diophantine provided, as usual, that D is a polynomial with integer coefficients. In Section 6.2 we used a form of Gödel coding, and knowledge of the values of the unknowns in (6.1.2) made it easy to find values of $x_1, \ldots, x_m$ for any value of y. In the construction described below, no such correspondence will be evident.

We begin by applying a technique similar to that used in Section 1.4 for isolating one variable. This will enable us to isolate the variable under the universal quantifier in (6.3.1); namely, we transform (6.3.1) into an equivalent formula of the form

$$\forall y < b\, \exists x_0 \ldots x_m\, [D_1(a, x_0, \ldots, x_m) = y], \tag{6.3.2}$$

where

$$D_1(a, x_0, \ldots, x_m) = (x_0 + 1)(1 - D^2(a, x_0, \ldots, x_m)) - 1. \tag{6.3.3}$$

Next, in a manner similar to that used in Section 6.2, we introduce an explicit bound on the values of $x_0, \ldots, x_m$; i.e., we rewrite (6.3.2) as the equivalent formula

$$\exists w\, \forall y < b\, \exists x_0 < w+1 \ldots \exists x_m < w+1\, [D_1(a, x_0, \ldots, x_m) = y]. \tag{6.3.4}$$

(The reason that a positive bound is needed will be explained later.)

As in Section 6.2, we require a polynomial $B(a, w)$ with natural number coefficients and with the property that the inequalities

$$x_0, \ldots, x_m \leq w \tag{6.3.5}$$

imply that

$$|D_1(a, x_0, \ldots, x_m)| < B(a, w). \tag{6.3.6}$$

In what follows we shall be working with two-sided estimates of some quantities, the upper and lower bounds being close one to another. Each such pair of bounds will have the form

$$A - B^- \leq X \leq A + B^+, \tag{6.3.7}$$

where A is the principal term and B^- and B^+ are small additional terms, whose exact forms will be of no importance to us. To simplify the notation and to emphasize the principal term, we will write the symbolic equation

$$X = A + \Theta(B) \tag{6.3.8}$$

instead of the pair of inequalities (6.3.7). Here, $\Theta(B)$ denotes an unspecified value in the interval $[-B, B]$ and may well have different values in different formulas. Readers familiar with the notation $O(T)$ will recognize this as a similar usage.

In Section 6.2, in order to combine an indefinite number of equations into a single condition, we transformed an equation into a congruence with a large modulus. In our present discussion, we shall make use of a less obvious transformation in which an important role will be played by the function unit(t), *the number of occurrences of the digit* 1 *in the binary representation of the number* t. Clearly, this function has the following property:

$$2^q > s \Longrightarrow \text{unit}(r) + \text{unit}(s) = \text{unit}(2^q r + s). \tag{6.3.9}$$

Suppose that the natural numbers p and q and integers α and β satisfy the inequalities

$$p > 2q, \qquad 2^q > |\alpha|, \qquad 2^q > |\beta|. \tag{6.3.10}$$

It is easy to check that

$$\text{unit}(2^p - 1 + \alpha - \beta) = \begin{cases} p + \Theta(q) & \text{if } \alpha \leq \beta, \\ \Theta(q) & \text{otherwise.} \end{cases} \tag{6.3.11}$$

Interchanging α and β and applying (6.3.9), we obtain

$$\text{unit}\big(2^{p+1}(2^p - 1 + \alpha - \beta) + 2^p - 1 - \alpha + \beta\big) = \begin{cases} 2p & \text{if } \alpha = \beta, \\ p + \Theta(2q) & \text{otherwise.} \end{cases} \tag{6.3.12}$$

We shall use this relationship with

$$\alpha = D_1(a, x_0, \ldots, x_m), \tag{6.3.13}$$

$$\beta = y, \tag{6.3.14}$$

$$q = B(a, w) + b + 1, \tag{6.3.15}$$

$$p = (b + 1)\big(4(w + 1)^{m+1} + 2\big)q. \tag{6.3.16}$$

Let us write

$$E(x_0, \dots, x_m, w) = (2^{p+1} - 1)D_1(a, x_0, \dots, x_m) + 2^{2p+1} - 2^p - 1, \tag{6.3.17}$$

$$F(w) = 2^{p+1} - 1. \tag{6.3.18}$$

(E and F also depend on a and b, but for the sake of simplicity of notation this dependence won't be indicated explicitly.) Using this notation, (6.3.12) becomes

$$\operatorname{unit}\Big(E(x_0, \dots, x_m, w) - F(w)y\Big) = \begin{cases} 2p & \text{if } D_1(a, x_0, \dots, x_m) = y, \\ p + \Theta(2q) & \text{otherwise.} \end{cases} \tag{6.3.19}$$

Thus, we have managed to construct a natural number, namely,

$$E(x_0, \dots, x_m, w) - F(w)y, \tag{6.3.20}$$

whose binary representation contains many or few 1s depending on whether or not $x_0, \dots, x_m$ and y satisfy the equation

$$D_1(a, x_0, \dots, x_m) = y. \tag{6.3.21}$$

Moreover,

$$0 \le E(x_0, \dots, x_m, w) - F(w)y < 2^{2p+2}, \tag{6.3.22}$$

and hence using (6.3.6), we can combine all $(w + 1)^{m+1}$ numbers (6.3.16), corresponding to different values of $x_0, \dots, x_m$ satisfying (6.3.5), into the single number

$$G(w, y) = K(w) - L(w)y, \tag{6.3.23}$$

where

$$K(w) = \sum_{x_0=0}^{w} \cdots \sum_{x_m=0}^{w} 2^{(2p+2)(x_m(w+1)^m + \cdots + x_0)} E(x_0, \dots, x_m, w), \tag{6.3.24}$$

$$L(w) = \sum_{x_0=0}^{w} \cdots \sum_{x_m=0}^{w} 2^{(2p+2)(x_m(w+1)^m + \cdots + x_0)} F(w). \tag{6.3.25}$$

By (6.3.19),

$$\operatorname{unit}(G(w, y)) = \left((w + 1)^{m+1} + s\right)p + \Theta(2(w + 1)^{m+1}q)\,, \tag{6.3.26}$$

where s is the number of m-tuples $x_0, \ldots, x_m$ satisfying (6.3.5) and (6.3.21). Thus, for equation (6.3.21) to have a solution with $x_0, \ldots, x_m$ satisfying inequalities (6.3.5), it is *necessary* that

$$\mathrm{unit}(G(w,y)) \geq (w+1)^{m+1}p + p - 2(w+1)^{m+1}q \tag{6.3.27}$$

and *sufficient* that

$$\mathrm{unit}(G(w,y)) > (w+1)^{m+1}p + 2(w+1)^{m+1}q. \tag{6.3.28}$$

By virtue of (6.3.16), inequality (6.3.27) is stronger than (6.3.28), and hence any inequality that is intermediate in power is both necessary and sufficient. For reasons that will become clear later, we choose as a necessary and sufficient condition the inequality

$$\mathrm{unit}(G(w,y)) \geq I(w) - J(w)y, \tag{6.3.29}$$

where

$$I(w) = (w+1)^{m+1}p + p - 2(w+1)^{m+1}q, \tag{6.3.30}$$

$$J(w) = (4(w+1)^{m+1} + 2)q. \tag{6.3.31}$$

Now, once again, we use Kummer's Theorem (see the Appendix), which provides the following representation for $\mathrm{unit}(t)$:

$$\mathrm{unit}(t) = \deg_2 \binom{2t}{t}, \tag{6.3.32}$$

where $\deg_2(r)$ is the exponent of 2 in the canonical representation of r as a product of primes.

Equation (6.3.32) enables us to eliminate the function unit from (6.3.29) by rewriting this condition in the form

$$2^{I(w)} \mid C(w,y), \tag{6.3.33}$$

where

$$C(w,y) = 2^{J(w)y} \binom{2G(w,y)}{G(w,y)}. \tag{6.3.34}$$

The left-hand side of (6.3.33) doesn't depend on y, so we can sum the right-hand side over y, obtaining

$$2^{I(w)} \mid S(w), \tag{6.3.35}$$

where

$$S(w) = \sum_{y=0}^{b-1} C(w, y). \tag{6.3.36}$$

It is obvious that the validity of (6.3.33) for $y = 0, \ldots, b-1$ implies the validity of (6.3.35). It is much less evident that *the single divisibility condition* (6.3.35) *relative to the sum* (6.3.36) *implies the corresponding divisibility condition* (6.3.33) *for each of the summands.* Let us verify that this is indeed the case.

Suppose that, on the contrary, (6.3.35) holds, but for some y less than b, (6.3.33) does not hold; let y_0 be the least such value of y. In other words, for y less than y_0 equation (6.3.21) has a solution satisfying (6.3.5), but for $y = y_0$ it does not.

By (6.3.34), (6.3.32), (6.3.28), (6.3.16), (6.3.30), and (6.3.31),

$$\begin{aligned} \deg_2(C(w, y_0)) &= J(w)y_0 + \text{unit}(G(w, y_0)) \\ &\leq J(w)y_0 + (w+1)^{m+1}p + 2(w+1)^{m+1}q \\ &< I(w). \end{aligned} \tag{6.3.37}$$

By (6.3.34), (6.3.32), (6.3.27), (6.3.30), and (6.3.37), for $y < y_0$,

$$\deg_2(C(w, y)) = J(w)y + \text{unit}(G(w, y)) \geq I(w) > \deg_2(C(w, y_0)). \tag{6.3.38}$$

By (6.3.34), (6.3.32), (6.3.26), (6.3.30), (6.3.31), and (6.3.37), for $y > y_0$,

$$\begin{aligned} \deg_2(C(w, y)) &= J(w)y + \text{unit}(G(w, y)) \\ &\geq J(w)(y_0 + 1) + (w+1)^{m+1}p - 2(w+1)^{m+1}q \\ &> \deg_2(C(w, y_0)). \end{aligned} \tag{6.3.39}$$

Thus, for $y \neq y_0$,

$$\deg_2(C(w, y)) > \deg_2(C(w, y_0)), \tag{6.3.40}$$

and hence

$$\deg_2(S(w)) = \deg_2(C(w, y_0)) < I(w), \tag{6.3.41}$$

which yields the desired contradiction with (6.3.35).

Thus, we have shown that formula (6.3.35) is equivalent to the formula

$$\forall y < b \, \exists x_0 < w+1 \ldots \exists x_m < w+1 \, [D_1(a, x_0, \ldots, x_m) = y], \tag{6.3.42}$$

and hence that formulas (6.3.4), (6.3.2), and (6.3.1) are equivalent to the formula

$$\exists w \big[2^{I(w)} \mid S(w) \big]. \tag{6.3.43}$$

This shows that we have eliminated the bounded universal quantifier, but at the cost of introducing summations with variable upper limits in (6.3.24), (6.3.25), and (6.3.36). Fortunately, all of these sums are easy to compute, and S turns out to be a Diophantine function.

The multiple sum in (6.3.25) is obviously equal to the product

$$\left(\sum_{x_0=0}^{w} 2^{(2p+2)x_0}\right)\cdots\left(\sum_{x_0=0}^{w} 2^{(2p+2)(w+1)^m x_m}\right)F(w), \qquad (6.3.44)$$

in which the first $m+1$ factors are sums of geometric progressions whose ratios are not equal to 1 (that was the reason for insisting on a positive bound in (6.3.4)). Hence,

$$\begin{aligned} L(w) &= \left(\frac{2^{(2p+2)(w+1)}-1}{2^{2p+2}-1}\right)\cdots\left(\frac{2^{(2p+2)(w+1)^{m+1}}-1}{2^{(2p+2)(w+1)^m}-1}\right)F(w) \\ &= \frac{L_1(w)-L_2(w)}{L_3(w)-L_4(w)}, \end{aligned} \qquad (6.3.45)$$

where $L_1(w)$, $L_2(w)$, $L_3(w)$, and $L_4(w)$ are suitable expressions constructed from particular natural numbers and the variables a, b, and w using addition, multiplication, and exponentiation, and such that for all a, b, and w, $L_3(w) > L_4(w)$.

By (6.3.17),

$$E(x_0,\ldots,x_m,w) = \sum_{i_0,\ldots,i_m} \lambda_{i_0,\ldots,i_m} x_0^{i_0}\ldots x_m^{i_m}, \qquad (6.3.46)$$

where the integer coefficients $\lambda_{i_0,\ldots,i_m}$ depend on a, b, and w. Substituting this expression for E into (6.3.24) and changing the order of summation, we see that

$$K(w) = \sum_{i_0,\ldots,i_m} \lambda_{i_0,\ldots,i_m} \sum_{x_0=0}^{w}\cdots\sum_{x_m=0}^{w} 2^{(2p+2)(x_m(w+1)^m+\cdots+x_0)} x_0^{i_0}\ldots x_m^{i_m}. \qquad (6.3.47)$$

The inner $(m+1)$-fold multiple sum in (6.3.47) is obviously equal to the product of sums of the form

$$\sum_{x=0}^{w} 2^{(2p+2)x(w+1)^k} x^{i_k}. \qquad (6.3.48)$$

Formulas for summing such generalized geometric progressions are provided in the

Appendix. Finally, we obtain for $K(w)$ an expression of the form

$$K(w) = \frac{K_1(w) - K_2(w)}{K_3(w) - K_4(w)}, \tag{6.3.49}$$

where analogously to (6.3.45), $K_1(w)$, $K_2(w)$, $K_3(w)$, and $K_4(w)$ are suitable expressions constructed from particular natural numbers and the variables a, b, and w by addition, multiplication, and exponentiation, and such that $K_3(w) > K_4(w)$ for all a, b, and w.

Now we turn to the sum (6.3.36). It contains $C(w, y)$, whose value was defined in (6.3.34) via binomial coefficients. These were shown to be Diophantine in Section 3.4 by constructing a positional code for tuple (3.4.1). Now, with u a sufficiently large number, we are going to construct a cipher to the base u of a tuple one of whose elements is equal to $S(w)$.

We have

$$(u+1)^{2G(w,y)} = \sum_{i=0}^{2G(w,y)} \binom{2G(w,y)}{i} u^i. \tag{6.3.50}$$

We are interested in the quantity

$$\binom{2G(w,y)}{G(w,y)}, \tag{6.3.51}$$

which occurs in (6.3.50) as the coefficient of $u^{G(w,y)} = u^{K(w)-L(w)y}$. Multiplying (6.3.50) by $2^{J(w)y}u^{L(w)y}$ and summing over y, we have:

$$\sum_{y=0}^{b-1} 2^{J(w)y} u^{L(w)y} (u+1)^{2K(w)-2L(w)y}$$

$$= \sum_{y=0}^{b-1} \sum_{i=0}^{2G(w,y)} 2^{J(w)y} \binom{2G(w,y)}{i} u^{i+L(w)y}. \tag{6.3.52}$$

Setting $u = 1$, we see that

$$\begin{aligned} \sum_{y=0}^{b-1} \sum_{i=0}^{2G(w,y)} 2^{J(w)y} \binom{2G(w,y)}{i} &= \sum_{y=0}^{b-1} 2^{2K(w)+(J(w)-2L(W))y} \\ &\leq 2^{2K(w)+J(w)b}(b+1). \end{aligned} \tag{6.3.53}$$

Hence, if

$$u > 2^{2K(w)+J(w)b}(b+1), \tag{6.3.54}$$

then the number (6.3.52) is the cipher to the base u of a tuple whose $(K(w)+1)$th element is equal to $S(w)$. The left-hand side in (6.3.52) is just the sum of an arithmetic progression, and we obtain

$$S(w) = \mathrm{Elem}((u+1)^{2K(w)}\left(\frac{(2^{J(w)}u^{L(w)}(u+1)^{2K(w)})^b - 1}{2^{J(w)}u^{L(w)}(u+1)^{2K(w)} - 1}\right), u, K(w)+1), \tag{6.3.55}$$

provided that (6.3.54) is satisfied.

Equations (6.3.44), (6.3.46), (6.3.52), (6.3.30), and (6.3.31) show that I, J, K, L, and S are Diophantine functions (of three arguments a, b, and w), and hence (6.3.43) can be viewed as a generalized Diophantine representation of the set defined by formula (6.3.1).

6.4 Connections between Hilbert's Eighth and Tenth Problems

The techniques developed in the previous sections of this chapter entitle us to use bounded universal quantifiers as a tool in constructing Diophantine representations. It is easy to see that the bound need not be a variable but can be the value of any Diophantine function. Formulas containing several bounded universal quantifiers can also be used to define Diophantine sets. The quantifiers can be eliminated one at a time, beginning with the innermost one.

Universal quantifiers, even if they are bounded, provide a very powerful tool for establishing that many different sets are Diophantine. For example, it is obvious that the set of primes is defined by the formula

$$a > 1 \;\&\; \forall x < a \forall y < a\, [a \neq (x+2)(y+2)], \tag{6.4.1}$$

which can now be transformed into a Diophantine representation of the set:

$$\mathrm{Prime}(a) \iff \exists x_1 \ldots x_m\, [\mathbf{P}(a, x_1, \ldots, x_m) = 0]. \tag{6.4.2}$$

In Section 2.5 we saw that among the individual subproblems of Hilbert's Tenth Problem, we can point to one in particular that is equivalent to Fermat's Last Theorem. Although it would not have been possible to do this prior to 1970, the fact in itself is not very surprising, because after all, Fermat's Last Theorem is also a statement about Diophantine equations. In this section we shall see that two other famous problems, seemingly having little to do with Diophantine equations, can nevertheless be reformulated as statements about the unsolvability of *particular* Diophantine equations.

We begin with *Goldbach's Conjecture*, which was included by Hilbert as part of his Eighth Problem. This still-unproved (1992) conjecture states that *every even number greater than 2 is the sum of two prime numbers.* With the polynomial $\mathbf{P}$ from (6.4.2) at our disposal, we can reformulate Goldbach's Conjecture as the statement that the Diophantine equation

$$(2a+4-p_1-p_2)^2+\mathbf{P}^2(p_1,x_1,\ldots,x_m)+\mathbf{P}^2(p_2,y_1,\ldots,y_m)=0 \quad (6.4.3)$$

is solvable in p_1, p_2, x_1, …, x_m, y_1, …, y_m for all values of the parameter a. This reformulation still doesn't make Goldbach's Conjecture an individual subproblem of Hilbert's Tenth Problem, because Hilbert asked for a method for determining the solvability of individual equations and did not speak of parametric families of equations.

To reduce Goldbach's Conjecture to a single equation, we note that the statement that the number $2a+4$ is *not* representable as the sum of two primes can easily be expressed using a bounded universal quantifier:

$$\forall z < a+1 \, \exists xy \, [z+2=(x+2)(y+2) \vee (2a+2-z)=(x+2)(y+2)]. \quad (6.4.4)$$

We can transform this formula into an equivalent Diophantine equation

$$\mathbf{G}(a,x_1,\ldots,x_m)=0 \quad (6.4.5)$$

that has a solution in x_1, …, x_m if and only if the value of the parameter a satisfies (6.4.4), i.e., refutes Goldbach's Conjecture. Replacing the parameter a by an unknown, we see that *Goldbach's Conjecture is equivalent to the statement that the equation*

$$\mathbf{G}(x_0,\ldots,x_m)=0 \quad (6.4.6)$$

has no solutions.

The Eighth Problem includes yet another famous conjecture also connected with primes and also still (1992) neither proved nor refuted, namely, the *Riemann Hypothesis.* This is a hypothesis about the complex zeros of *Riemann's zeta function*, which can be defined for $\mathrm{Re}(\alpha) > 1$ by

$$\zeta(\alpha)=\sum_{n=1}^{\infty}\frac{1}{n^{\alpha}}. \quad (6.4.7)$$

A connection of the zeta function with primes can be seen from *Euler's Identity*

$$\sum_{n=1}^{\infty} \frac{1}{n^{\alpha}} = \prod_{p} \left(1 - \frac{1}{p^{\alpha}}\right)^{-1}, \tag{6.4.8}$$

in which the product in the right-hand side is taken over all primes. The expression under the product symbol in (6.4.8) can be rewritten as

$$\left(1 - \frac{1}{p^{\alpha}}\right)^{-1} = 1 + \frac{1}{p^{\alpha}} + \frac{1}{p^{2\alpha}} + \cdots, \tag{6.4.9}$$

and then it becomes clear that identity (6.4.8) is an analytic form of the Fundamental Theorem of Arithmetic.

Both the series and the product in (6.4.8) converge only for $\text{Re}(\alpha) > 1$ and define an analytical function on this half-plane, which can be extended using analytic continuation to the entire complex plane except for the point $\alpha = 1$, which is a simple pole of the zeta function. It is known that

$$0 = \zeta(-2) = \zeta(-4) = \zeta(-6) = \cdots, \tag{6.4.10}$$

and the numbers -2, -4, -6, ... are called the *trivial zeros* of the zeta function. The Riemann Hypothesis states that all the other, *non-trivial* complex zeros lie on the line $\text{Re}(\alpha) = 1/2$.

The Riemann Hypothesis plays an important role in research on the distribution of the primes in the set of natural numbers. Let $\pi(n)$ denote, as usual, the number of primes $\leq n$. According to the *Prime Number Theorem*,

$$\pi(n) = \frac{n}{\ln(n)}(1 + o(1)). \tag{6.4.11}$$

The elementary function $n/\ln(n)$ is in fact not a very good approximation to $\pi(n)$; a much better approximation is given by the non-elementary function

$$\int_{2}^{n} \frac{d\alpha}{\ln(\alpha)}. \tag{6.4.12}$$

The Riemann Hypothesis can be reformulated as the statement

$$\pi(n) = \int_{2}^{n} \frac{d\alpha}{\ln(\alpha)} + +O(\sqrt{n}\ln(n)). \tag{6.4.13}$$

For technical reasons, it is more convenient to work not with the function π, but with a related function, *Chebyshev's* ψ, which can be defined by

$$\psi(n) = \ln(\text{lcm}(1, 2, \ldots, n)). \tag{6.4.14}$$

While the function π jumps by 1 at every prime p, the function ψ jumps by $\ln(p)$ at p itself and at all of its powers. In terms of the function ψ, the Prime Number Theorem can be written as

$$\psi(n) = n + o(n). \tag{6.4.15}$$

Here the principal term n is a very good approximation, and the Riemann Hypothesis is equivalent to the statement

$$\psi(n) = n + O(\sqrt{n}\ln^2(n)). \tag{6.4.16}$$

In order to be able to reformulate the Riemann Hypothesis as a statement about a particular equation, we will need the value of an actual numerical constant in the error term, rather than just what is implied by the symbol O. In fact, it is known that the Riemann Hypothesis implies that

$$|\psi(n) - n| < \sqrt{n}\ln^2(n) \tag{6.4.17}$$

for

$$n \geq 600. \tag{6.4.18}$$

The function lcm occurring in (6.4.14) was shown to be Diophantine in Section 1.6, but only when considered as a function of two arguments. A bounded universal quantifier enables us to cope with the indefinite numbers of arguments of lcm in (6.4.14). Namely, the number m is a common multiple of the numbers $1, \dots, n$ if and only if

$$\forall y < n\,[(y+1) \mid m] \tag{6.4.19}$$

and is the least common multiple if, in addition,

$$m > 0 \,\&\, \forall y < m\big[y = 0 \vee \exists x < n\,[(x+1) \nmid y]\big]. \tag{6.4.20}$$

We cannot discuss the question of whether the function ln, also occurring in (6.4.14), is or is not Diophantine because its values are not, in general, integers. We could show that the function $\lfloor \ln(n) \rfloor$ (where $\lfloor \alpha \rfloor$ denotes the integer part of α) is Diophantine, but that would require the use of some quite nontrivial fact such as the irrationality of the base of natural logarithms. (In following the usual convention of denoting this number by e, we are permitting ourselves to violate our convention of using lower-case italic Latin letters only for natural numbers.)

Instead of the function ln, we introduce the relation

$$\operatorname{explog}(a,b) \iff \exists x \left[x > b+1 \;\&\; \left(1+\frac{1}{x}\right)^{xb} \le a+1 < 4\left(1+\frac{1}{x}\right)^{xb} \right] \tag{6.4.21}$$

and show, first, that for all a and b,

$$\operatorname{explog}(a,b) \Longrightarrow |b - \ln(a+1)| < 2 \tag{6.4.22}$$

and, second, that

$$\forall a\, \exists b\, [\operatorname{explog}(a,b)]. \tag{6.4.23}$$

In fact, it is known from calculus that

$$e = \lim_{x\to\infty}\left(1+\frac{1}{x}\right)^{x} = \lim_{x\to\infty}\left(1+\frac{1}{x}\right)^{x+1}, \tag{6.4.24}$$

and, moreover, that for $x > 0$,

$$\left(1+\frac{1}{x}\right)^{x} < e < \left(1+\frac{1}{x}\right)^{x+1}. \tag{6.4.25}$$

Raising this last inequality to the power b, we get

$$\left(1+\frac{1}{x}\right)^{xb} < e^b < \left(1+\frac{1}{x}\right)^{(x+1)b} \tag{6.4.26}$$

and hence

$$\frac{e^b}{e} \le \frac{\left(1+\frac{1}{x}\right)^{(x+1)b}}{\left(1+\frac{1}{x}\right)^{x}} < \left(1+\frac{1}{x}\right)^{xb} \le a+1 < 4\left(1+\frac{1}{x}\right)^{xb} \le 4e^b, \tag{6.4.27}$$

which implies the inequality in (6.4.22).

On the other hand, for any a one can find b such that

$$b \le \ln(a+1) < b + \ln(3). \tag{6.4.28}$$

Choosing x very large, we can, by (6.4.24), make both of the quantities

$$\left(1+\frac{1}{x}\right)^{x}, \quad \left(1+\frac{1}{x}\right)^{x+1} \tag{6.4.29}$$

arbitrarily close to e and thus guarantee the validity of the inequalities in (6.4.21).

We now show that *the negation of the Riemann Hypothesis is equivalent to the existence of numbers k, l, m, and n satisfying conditions* (6.4.18)–(6.4.20) *and*

$$\text{explog}(m-1,l), \tag{6.4.30}$$
$$\text{explog}(n-1,k), \tag{6.4.31}$$
$$(l-n)^2 > 4n^2k^4. \tag{6.4.32}$$

First, suppose that there are such numbers k, l, m, and n. Then by (6.4.19) and (6.4.20),

$$m = e^{\psi(n)}, \tag{6.4.33}$$

and by (6.4.22) it follows from (6.4.31) that

$$|l-\psi(n)| < 2. \tag{6.4.34}$$

Similarly, it follows from (6.4.32) that

$$|k-\ln(n)| < 2. \tag{6.4.35}$$

Thus, we obtain

$$|\psi(n)-n| \geq |l-n| - |l-\psi(n)| > 2\sqrt{n}k^2 - 2 > \sqrt{n}\ln^2(n), \tag{6.4.36}$$

which gives the desired contradiction with (6.4.17).

Now suppose that the Riemann Hypothesis is not true. Then by (6.4.16) there is a number n satisfying (6.4.18) such that

$$|\psi(n)-n| > 10\sqrt{n}\ln^2(n). \tag{6.4.37}$$

Choosing m according to (6.4.33), conditions (6.4.19) and (6.4.20) will be satisfied. Then by (6.4.23) we can find numbers k and l satisfying (6.4.30) and (6.4.31). By (6.4.22) the inequalities (6.4.34) and (6.4.35) will be satisfied. They in turn imply that

$$\begin{aligned} |l-n| &\geq |\psi(n)-n| - |l-\psi(n)| \\ &> 10\sqrt{n}\ln^2(n) - 2 \\ &> 2\sqrt{n}\big(\ln(n)+2\big)^2 \\ &> 2\sqrt{n}k^2, \end{aligned} \tag{6.4.38}$$

i.e., that inequality (6.4.32) is also satisfied.

Taking advantage of the fact that all of the conditions (6.4.18)–(6.4.20), (6.4.30)–(6.4.32) are Diophantine, we can construct a Diophantine equation whose unsolvability is equivalent to the Riemann Hypothesis.

It would be a mistake to expect that every problem can be reduced directly to a question concerning the solvability of Diophantine equations. For example, since we are able to eliminate only *bounded* universal quantifiers, we have no obvious way to restate the twin prime conjecture (also included by Hilbert in his Eighth Problem) as the problem of the solvability or unsolvability of a particular Diophantine equation.

6.5 Yet another universal equation

In this section we shall see how bounded universal quantifiers allow one to easily construct another universal Diophantine equation.

The coding of Diophantine equations will be different from those used in Chapter 4. It is natural that prior to coding equations, we introduce numbering of polynomials. Without loss of generality, we will number only polynomials in the variables x_1, x_2, ... We define a *universal sequence of polynomials* as follows:

$$\begin{aligned} \mathbf{P}_{4k} &= k, \\ \mathbf{P}_{4k+1} &= \mathbf{P}_i + \mathbf{P}_j, \\ \mathbf{P}_{4k+2} &= \mathbf{P}_i \mathbf{P}_j, \\ \mathbf{P}_{4k+3} &= x_{k+1}, \end{aligned} \tag{6.5.1}$$

where

$$k = \text{Cantor}(i, j). \tag{6.5.2}$$

(This definition is correct because the polynomial Cantor, introduced in Section 3.1, maps the pairs of natural numbers one-to-one to the natural numbers.)

Clearly, the sequence

$$\mathbf{P}_0, \mathbf{P}_1, \ldots \tag{6.5.3}$$

is universal in the following sense: *every polynomial with natural number coefficients in the variables x_1, x_2, ... occurs in* (6.5.3).

A tuple of natural numbers

$$\langle p_0, p_1, \ldots, p_{r-1} \rangle \tag{6.5.4}$$

will be called a *realization* of the sequence (6.5.3) if for some choice of values for the variables x_1, x_2, ...,

$$p_k = \mathbf{P}_k(x_1, x_2, \dots) \tag{6.5.5}$$

for $k = 0, 1, \dots, r-1$.

We have two techniques available for coding tuples of indefinite length, namely, Gödel coding, introduced in Section 3.2, and positional coding, introduced in Section 3.3. While either could be used with equal success, it is more natural to use Gödel coding because it is much easier to prove that the function GElem introduced in (3.2.15) is Diophantine than to do the same for the function Elem introduced in (3.3.8).

The relation *tuple* $\langle p, q, r\rangle$ *is a code of a realization*, written $\mathrm{Real}(p, q, r)$, can easily be expressed using a bounded universal quantifier:

$$\begin{aligned}
&\mathrm{Real}(p,q,r) \iff \\
&\forall l < r\, \exists ijk \Big[k = \mathrm{Cantor}(i,j) \,\&\, \big[[l = 4k \,\&\, \mathrm{GElem}(p,q,l+1) = k] \\
&\quad \vee [l = 4k+1 \,\&\, \mathrm{GElem}(p,q,l+1) = \mathrm{GElem}(p,q,i+1) + \mathrm{GElem}(p,q,j+1)] \\
&\quad \vee [l = 4k+2 \,\&\, \mathrm{GElem}(p,q,l+1) = \mathrm{GElem}(p,q,i+1)\,\mathrm{GElem}(p,q,j+1)]\big]\Big].
\end{aligned} \tag{6.5.6}$$

The number of the equation

$$P_i(a_1, \dots, a_n, x_{n+1}, \dots) = P_j(a_1, \dots, a_n, x_{n+1}, \dots) \tag{6.5.7}$$

is defined to be $\mathrm{Cantor}(i, j)$. Clearly, the equation whose number is k has a solution if and only if the parameters a_1, ..., a_n satisfy the condition

$$\begin{aligned}
\exists ijpqr\, [k = \mathrm{Cantor}(i,j) \,&\&\, r > i \,\&\, r > j \\
&\&\, \mathrm{Real}(p,q,r) \,\&\, \mathrm{GElem}(p,q,i+1) = \mathrm{GElem}(p,q,j+1) \\
&\&\, \mathrm{GElem}(p,q,5) = a_1 \,\&\, \cdots \,\&\, \mathrm{GElem}(p,q,4n+1) = a_n].
\end{aligned} \tag{6.5.8}$$

This condition is evidently Diophantine and hence can be transformed into the desired universal equation.

6.6 Yet another Diophantine set with non-Diophantine complement

In Section 10.1 we shall need the existence of a Diophantine set $\mathfrak{S}$ of natural numbers whose complement $\overline{\mathfrak{S}}$ is not Diophantine and that has an additional special property. Now, $\overline{\mathfrak{S}}$ being non-Diophantine means that for any Diophantine set $\mathfrak{D}$, at least one of the following assertions is valid:

$$\text{there exists } a \text{ such that } a \in \mathfrak{D}, \text{ but } a \notin \overline{\mathfrak{S}}, \tag{6.6.1}$$

$$\text{there exists } a \text{ such that } a \notin \mathfrak{D}, \text{ but } a \in \overline{\mathfrak{S}}. \tag{6.6.2}$$

Clearly, the set $\overline{\mathfrak{S}}$ must be infinite; hence the assertion (6.6.2) will be true for any finite set $\mathfrak{D}$. It turns out that the set $\mathfrak{S}$ can be defined in such a way that for any infinite $\mathfrak{D}$, assertion (6.6.1) will be valid. In other words, the set $\mathfrak{S}$ will contain a representative from each infinite Diophantine set. A Diophantine set $\mathfrak{S}$ such that (6.6.1) is true for every infinite Diophantine set and (6.6.2) is true for every finite set $\mathfrak{D}$ is called *simple*.

In order to construct a simple set $\mathfrak{S}$, we begin with a universal Diophantine equation

$$U(a, k, x_1, \ldots, x_m) = 0. \tag{6.6.3}$$

(At this point, two methods for constructing universal equations have been presented, one in the previous section and one in Section 4.5.) We define the set $\mathfrak{S}$ by

$$\begin{aligned} \exists k \Big[&\exists x_1 \ldots x_m \,[2k < a \,\&\, U(a, k, x_1, \ldots, x_m) = 0] \\ &\& \,\forall y < \text{Cantor}_{m+1}(a, x_1, \ldots, x_m) \,\exists z_0 \ldots z_m \\ &\quad [y = \text{Cantor}_{m+1}(z_0, \ldots, z_m) \,\&\, [2k \geq z_0 \vee U(z_0, k, z_1, \ldots, z_m) \neq 0]]\Big]. \end{aligned} \tag{6.6.4}$$

Let us check conditions (6.6.1) and (6.6.2). Let $\mathfrak{D}$ be an infinite Diophantine set; then there is a number k such that $\mathfrak{D}$ is defined by the equation

$$U(a, k, x_1, \ldots, x_m) = 0. \tag{6.6.5}$$

Because $\mathfrak{D}$ is infinite, there are numbers a, x_1, $\ldots$, x_m satisfying (6.6.5) as well as the inequality

$$2k < a. \tag{6.6.6}$$

Among all of these $(m+1)$-tuples, we chose the one with the least Cantor number in (6.6.4). It is not difficult to see that the initial element of this tuple can play the role of a in (6.6.1).

Now let $\mathfrak{D}$ be a finite set with l elements d_1, $\ldots$, d_l. Suppose that (6.6.2) is not true, i.e., that every a different from d_1, $\ldots$, d_l belongs to the set $\mathfrak{S}$. For every a

in $\mathfrak{S}$, let k_a denote the corresponding number whose existence is asserted in (6.6.4). In particular, among the numbers 0, 1, ..., $2l$ there are $l+1$ numbers $a_0, \ldots, a_l$ belonging to $\mathfrak{S}$, and hence there must exist corresponding numbers $k_{a_0}, \ldots, k_{a_l}$. It is not difficult to see that they must all be different and yet satisfy the inequalities

$$k_{a_i} < \frac{a_i}{2} < l, \tag{6.6.7}$$

which is impossible. This contradiction proves (6.6.2).

Exercises

1. Show that a bounded universal quantifier can be eliminated using moduli chosen according to (3.2.7) rather than according to (6.2.12).

2. Prove the following multiplicative version of Dirichlet's Principle: if $q \mid s_1 \cdots s_n$, then there are numbers r and k such that $r \mid q,\ 1 \le k \le n,\ r \mid s_k$, and $r^n \ge q$.

3. Generalize the previous exercise in the following manner: if $t \ne 0$ and

$$q \mid ts_{1,1} \cdots s_{1,n_1}, \ \ldots \ , q \mid ts_{m,1} \cdots s_{m,n_m},$$

then there are numbers r and $k_1, \ldots, k_m$ such that

$$r \mid q, \qquad 1 \le k_i \le n_i, \qquad r \mid s_{i,k_i} \qquad tr^{n_1 \cdots n_m} \ge q.$$

4. Using the previous exercise, show that condition (6.2.24) can be replaced by a sufficiently strong inequality of the form $q > \cdots$, which would then eliminate the need to prove that the factorial is Diophantine.

5. In addition to (6.2.13) and the product from Exercise 3.3, there are other cases when it is possible to give a direct proof that a product with indefinite upper limit is Diophantine. Check that

$$\alpha_b(2n) = \big(\alpha_b(n+1) - \alpha_b(n-1)\big)\alpha_b(n)$$

(where the sequences α_b are the ones defined in equation (2.1.4)) and hence

$$\prod_{k=1}^{m} \big(\alpha_b(2^k+1) - \alpha_b(2^k-1)\big) = \alpha_b(2^{m+1}),$$

and use the latter identity instead of (6.2.13) in order to eliminate bounded universal quantifiers.

6. Show that the Four Color Problem is yet another example of a famous problem that can be restated as an assertion of the unsolvability of a particular Diophantine equation.

7. Show that the set of numbers of unsolvable parameter-free Diophantine equations is not Diophantine.

8. A property $\mathcal{R}(a_1, \ldots, a_n)$ is called *polynomial* if it has a Diophantine representation without unknowns, i.e.,

$$\mathcal{R}(a_1, \ldots, a_n) \iff R(a_1, \ldots, a_n) = 0,$$

where R is a polynomial with integer coefficients. Show that if $\mathcal{R}$ is a polynomial relation, then so is the relation defined by formula (6.1.1).

Commentary

It has already been explained in the Commentary to Chapter 5 that the elimination of bounded universal quantifiers was an essential step in the original proof of the unsolvability of Hilbert's Tenth Problem. The appropriate technique was introduced by Davis, Putnam, and Robinson [1961]. In Section 6.2 we followed this work with some minor simplifications. More radical modifications were proposed by Matiyasevich [1972a, 1973] (see Exercise 6.4) and by Hirose and Iida [1973] (see Exercise 6.5).

The elimination of bounded universal quantifiers by the method of Section 6.1 became possible only after a direct method of simulating Turing machines by Diophantine equations was provided by Matiyasevich [1976].

The third construction given in Section 6.3 is introduced here for the first time. It is a development of the technique that was originally proposed by Matiyasevich [1979] for reducing the number of unknowns in exponential Diophantine equations (see Section 8.2). The novelty as compared to that paper is the compression of the system of divisibilities (6.3.33) into a single divisibility (6.3.35).

The possibility of translating many problems into the language of arithmetic was already demonstrated by Gödel [1931], and once it became possible to eliminate bounded universal quantifiers, reformulation in the language of Diophantine equations became a trivial matter. Perhaps the Riemann Hypothesis, dealing as it does with the continuum of real numbers, is an exception in this respect. Turing [1939] showed that the hypothesis is equivalent to a statement of the form

$$\forall y \, \exists x \, \mathcal{T}(x, y), \tag{1}$$

where $\mathcal{T}$ is a decidable relation between natural numbers x and y. However, this result does not suffice to reduce the negation of the Riemann Hypothesis to the unsolvability of a particular Diophantine equation, because the universal quantifier in (1) is not bounded. An improvement of Turing's result follows from Kreisel's paper [1958], which contains very general results about the arithmetization of assertions concerning analytic functions, in particular that the Riemann Hypothesis can be stated in the form

$$\forall y \, \mathcal{K}(y), \tag{2}$$

where $\mathcal{K}$ is a decidable property. The negation of $\mathcal{K}$ is semidecidable and hence Diophantine, so

$$\neg\mathcal{K}(y) \iff \exists x_1 \dots x_m \, [\mathbf{K}(y, x_1, \dots, x_m) = 0]; \tag{3}$$

thus, the Riemann Hypothesis is equivalent to the unsolvability of the equation

$$\mathbf{K}(x_0, \dots, x_m) = 0. \tag{4}$$

Kreisel's constructions are very general and would result in a very complicated equation. A more direct method, based on the connection between the Riemann Hypothesis and the distribution of prime numbers, was presented by Davis, Matiyasevich, and Robinson [1976]. The presentation in Section 6.4 includes further simplifications proposed to the author by Matti Jutila and Andrzej Schinzel. The simplification results from using the function ψ itself rather than the integral

$$\psi_1(n) = \int_1^n \psi(\alpha) \, d\alpha. \tag{5}$$

The function ψ_1 was used by Davis, Matiyasevich, and Robinson [1976] because of the need to have a definite value for the constant implied by the symbol O. We now have such a constant for ψ in (6.4.21) due to the work of Schoenfeld [1976].

In Chapter 5 we proved the undecidability of Hilbert's Tenth Problem; hence reductions of various problems to Diophantine equations provide little hope that these problems might be solved via analysis of the corresponding equations. The inverse view of such reductions may well turn out to be more fruitful: for example, the solution of the Four Color Problem can be presented as a very non-trivial method for proving the unsolvability of the corresponding Diophantine equation (see Exercise 6.6).

The numbering of polynomials in Section 6.5, based on building them up from variables and constant natural numbers using addition and multiplication, was given in Robinson [1969b] and used to construct a universal equation in Robinson [1971].

The notion of simple set was introduced by Post, and an example of a simple set was given in Post [1944]. (Of course, Post's construction was in terms of semidecidable sets, the equivalence with Diophantine sets not being known at the time.)

7 Decision Problems in Number Theory

Because of the great simplicity of polynomials with integer coefficients as mathematical objects, Hilbert's Tenth Problem has turned out to be a convenient tool for establishing the algorithmic unsolvability of various other decision problems. In this chapter we consider some examples of this kind from number theory.

7.1 The number of solutions of Diophantine equations

For every Diophantine equation

$$D(x_1, \ldots, x_m) = 0, \tag{7.1.1}$$

there is an associated quantity Card(D), the number of solutions of this equation. The values of Card belong to the set $\mathfrak{N} = \{0, 1, \ldots, \infty\}$. For every set $\mathfrak{A}$ such that $\mathfrak{A} \subseteq \mathfrak{N}$, one can consider the set $\text{Card}^{-1}(\mathfrak{A})$, the inverse image of $\mathfrak{A}$. Taking some liberty with the language, we will consider $\text{Card}^{-1}(\mathfrak{A})$ both as a set of equations and as a set of natural numbers, the codes of these equations.

The set

$$\text{Card}^{-1}(\{1, 2, \ldots, \infty\}) \tag{7.1.2}$$

was already considered in Section 4.6, since it is just the set of equations that have solutions, and Hilbert's Tenth Problem is precisely the problem of identifying the elements of the set (7.1.2) within the set of all equations. It was shown in Chapter 5 that the set (7.1.2) is undecidable (in the intuitive sense). To make this statement more precise, we can consider the set (7.1.2) as a set of codes and say that it is Turing undecidable. According to Section 5.6, this is equivalent to the statement that either the set (7.1.2) itself or its complement is not Diophantine, and in fact in Section 4.6 we found that the set (7.1.2) is Diophantine, while its complement, the set $\text{Card}^{-1}(\{0\})$, is not.

In this section we generalize this result by showing that *if* $\text{Card}^{-1}(\mathfrak{A})$ *is Diophantine, then either* $\mathfrak{A}$ *is empty or* $\infty \in \mathfrak{A}$. In particular, neither $\text{Card}^{-1}(\{1\})$, the set of equations with a unique solution, nor $\text{Card}^{-1}(\{0, 1, \ldots\})$, the set of equations with a finite number of solutions, is semidecidable.

Since

$$\overline{\text{Card}^{-1}(\mathfrak{A})} = \text{Card}^{-1}(\overline{\mathfrak{A}}), \tag{7.1.3}$$

with the exception of the two extreme cases $\mathfrak{A} = \emptyset$ *and* $\mathfrak{A} = \mathfrak{N}$, *the set* $\text{Card}^{-1}(\mathfrak{A})$ *is undecidable*, because either it or its complement does not contain ∞ and hence is not semidecidable. This statement can be rephrased in the following way: if the answer to a question about whether the number of solutions of a Diophantine equation belongs to some particular set will be positive for some equations and negative for others, then the answer to that question for an arbitrary given equation cannot be found algorithmically.

We limit ourselves to a semiformal proof based on Church's Thesis; namely, we will deal with semidecidability in the intuitive sense. The proof will be by contradiction; i.e., we suppose that for some non-empty set $\mathfrak{A}$ not containing ∞, the set $\mathrm{Card}^{-1}(\mathfrak{A})$ is semidecidable. We are going to show that in that case, the set $\mathrm{Card}^{-1}(\{0\})$ would also be semidecidable.

Let (7.1.1) be an arbitrary equation. Let us fix an element a from $\mathfrak{A}$ and consider the equation

$$\big((x_1 + x_2 + 1 - a)^2 + x_3^2 + \cdots + x_{m+1}^2\big) D(x_1, \ldots, x_m) = 0. \qquad (7.1.4)$$

It is easy to see that if equation (7.1.1) has no solution, then equation (7.1.4) has exactly a solutions

$$\begin{array}{lll} x_1 = 0, & x_2 = a - 1, & x_3 = \cdots = x_{m+1} = 0; \\ x_1 = 1, & x_2 = a - 2, & x_3 = \cdots = x_{m+1} = 0; \\ & \vdots & \\ x_1 = a - 1, & x_2 = 0, & x_3 = \cdots = x_{m+1} = 0. \end{array} \qquad (7.1.5)$$

On the other hand, if equation (7.1.1) has a solution, then equation (7.1.4) has infinitely many solutions, because x_{m+1} can be chosen arbitrarily. Thus equation (7.1.1) belongs to $\mathrm{Card}^{-1}(\{0\})$ if and only if equation (7.1.4) belongs to $\mathrm{Card}^{-1}(\mathfrak{A})$. So, a hypothetical method that semidecides $\mathrm{Card}^{-1}(\mathfrak{A})$ would allow us to semidecide $\mathrm{Card}^{-1}(\{0\})$ as well, which is impossible, as was noted above. We have obtained the desired contradiction.

7.2 Non-effectivizable estimates in the theory of exponential Diophantine equations

In Chapter 2 we established that exponentiation is Diophantine, thus obtaining a method for transforming an arbitrary exponential Diophantine equation into an equivalent, with respect to solvability, ordinary Diophantine equation. It might seem that every result concerning exponential Diophantine equations could be transferred to the ordinary Diophantine case. However, this is not true. In this section we present a result about exponential Diophantine equations that cannot at this time be strengthened to a similar result for ordinary Diophantine equations.

In number theory there are many theorems stating that Diophantine equations of some special kind have only finitely many solutions. As a classical example, we can mention the theorem due to Axel Thue [1909] stating that for every value of the parameter a, the equation

$$F(x, y) = a \tag{7.2.1}$$

has only finitely many solutions provided that F is an irreducible binary form of degree at least three. Thus, for every value of a there exists a number b such that every solution of equation (7.2.1) satisfies the inequalities $x < b, \quad y < b$. However, Thue's proof was by contradiction and did not yield a value for b. In such a situation, one speaks of a *non-effective* estimate for the size of solutions.

When, for some class of equations, a non-effective estimate for the size of solutions is obtained, the problem of effectivization naturally arises, i.e., the problem of finding an effectively calculable function of the parameters that would bound from above the maximum possible value of an unknown. If we have such an effective bound, we can, at least in principle, find all the solutions (or establish that there are none) by surveying all possible combinations of the values of the unknowns not exceeding the bound. In the case of Thue's Theorem, such effectivization was achieved only half a century later by Alan Baker [1968] and required radically new ideas. For many other classes of Diophantine equations, we still only have non-effective estimates.

The aim of this section is to show that there may be fundamental algorithmical obstacles to effectivization. So far it has been possible to prove this only for estimates of solutions of exponential Diophantine equations. Namely, we shall show that *there is an exponential Diophantine equation*

$$E(a, x_1, \ldots, x_m) = 0 \tag{7.2.2}$$

that, for each value of the parameter a, has at most one solution, but such that this unique solution cannot be bounded from above by a function whose values can be effectively calculated for every value of the parameter a.

In order to state this result rigorously, it would be necessary to give a formal definition of *effectively calculable function.* This could be done on the basis of the notion of Turing machine (see Exercise 5.5), but we content ourselves with a reference to Church's Thesis, using the connection, mentioned above, between the existence of effectively calculable estimates and the possibility of determining whether or not solutions exist. In fact, it is not difficult to see that we can take any equation defining an undecidable set and having at most one solution for each value of the parameter a as the exponential Diophantine equation (7.2.2).

This suggests the following definition: a set $\mathfrak{M}$ of n-tuples of natural numbers is said to have the *singlefold* representation

$$\langle a_1, \ldots, a_n \rangle \in \mathfrak{M} \iff \exists x_1 \ldots x_m \, [E(a_1, \ldots, a_n, x_1, \ldots, x_m) = 0] \quad (7.2.3)$$

if for every tuple of values of the parameters, the equation has at most one solution. Depending on the type of equation, one speaks of a *singlefold Diophantine* or of a *singlefold exponential Diophantine* representation. In a similar manner, one can speak of a singlefold or non-singlefold representation of a property, relation, or function.

The main result of this section is as follows: *every Diophantine set has a singlefold exponential Diophantine representation.* At present we cannot improve this result to obtain a singlefold Diophantine representation, because the technique of Chapter 2 provides only, so to speak, an infinity-fold Diophantine representation for exponentiation.

So, let us choose some Diophantine set $\mathfrak{M}$. Below we describe three techniques for finding a singlefold Diophantine representation of $\mathfrak{M}$.

First technique. The first technique is based on that of Chapter 5. Given a Diophantine representation of a set $\mathfrak{M}$, we can, following Section 5.4, find a Turing machine M semideciding $\mathfrak{M}$ and then, following Section 5.5, construct from this machine an exponential Diophantine representation of $\mathfrak{M}$. The machine M, when searching for a solution of the equation defining the set $\mathfrak{M}$, examines possible solutions in the order of their Cantor number and halts as soon as it finds the solution with the least number. Figuratively speaking, the machine does not "see" the remaining possible solutions with larger Cantor numbers. By modifying slightly the technique of Section 5.5, we shall obtain an exponential Diophantine representation whose unique solution corresponds to the least solution of the given Diophantine equation.

The modification of the technique for simulating Turing machines by Diophantine equations must begin with Chapter 1.

If (1.4.3) and (1.4.4) are singlefold representations of two sets, then (1.4.6) is a singlefold representation of their intersection. Thus, we are free to use conjunction in generalized singlefold representations. On the other hand, it is evident that

(1.4.5) need not be a singlefold representation for the union, so we have to avoid disjunction if we want to obtain a singlefold representation.

The representations (1.5.1), (1.6.1), and (1.6.2) evidently are singlefold, so the relations $\neq$, $\leq$, and $<$ remain in our arsenal. The representation (1.6.3) is not singlefold since, as defined, the divisibility relation is true for $a = b = 0$, and in that case x may have any arbitrary value. One way to deal with this is to always make sure that the first argument of the relation $|$ is not equal to zero. Alternatively, we can replace (1.6.3) by the generalized singlefold representation

$$a \mid b \iff \exists x\, [ax = b \,\&\, x \leq b]. \tag{7.2.4}$$

Either way, we can consider (1.6.4) to be a generalized singlefold representation of the function rem. This, in turn, implies that representations (1.6.7)–(1.6.9) are singlefold, so we are free to use the relation $|$, the function div, and congruences.

We don't need to change anything in Chapter 2 because we are concerned with exponential Diophantine representations.

In Chapter 3 we only need to revise the sections dealing with positional coding.

Formula (3.3.7) is evidently a singlefold exponential Diophantine representation for the relation Code. It is not difficult to convince oneself that the representation (3.3.8) of the function Elem is singlefold.

Equivalence (3.3.9) is not a representation for the nine-place relation Concat because in the left-hand side the second, fifth, and eighth arguments are the same. Representation (3.5.15) of this relation clearly is not singlefold. However, we do not need a singlefold representation for Concat as a nine-place relation because in Section 5.5 we deal with codes to a fixed base, and so the operation $+$ defined on such codes will suffice for our purposes. Formula (3.3.9) justifies the use of this operation for constructing singlefold exponential Diophantine representations.

Clearly, formulas (3.4.2) and (3.4.6) are singlefold representations for binomial coefficients and the factorial. It would not be difficult to modify representation (1.6.10) for the greatest common divisor to render it singlefold and hence to do the same for representation (3.4.7) of the primes. However, we prefer a simpler method. The fact that the primes are Diophantine was required only for using Kummer's Theorem to obtain the Diophantine representations (3.5.6) for PNotGreater and (3.5.8) for PSmall and then (3.5.13) for Eq, (3.5.4) for Equal, (3.5.13) for NotGreater, and (3.5.14) for Small. In Chapter 5 we work with a fixed base that is required only to be greater than the number of states of the Turing machine and the number of tape symbols. Now, in addition, we require the base to be prime. Thus, defining the binary relation PNotGreater_p by fixing the third argument in the ternary relation

PNotGreater equal to a prime p, it will suffice to give the singlefold exponential Diophantine representation

$$\text{PNotGreater}_p(a_1, a_2) \iff p \mid \binom{a_2}{a_1}. \tag{7.2.5}$$

An analog for (3.5.8) is the singlefold representation

$$\text{PSmall}_p(a, c, e) \iff \text{PNotGreater}_p(a, \text{Repeat}(e, p, c)) \tag{7.2.6}$$

for the relation $\text{PSmall}_p(a, c, e)$ which coincides with $\text{PSmall}(a, p, c, e)$ when p is prime and $e < p$. This is sufficient for obtaining the following analog for the representation (3.6.8):

$$\text{Ort}_p(q_1, q_2, r) \iff \text{PSmall}_p(q_1, r, 1) \,\&\, \text{PSmall}_p(q_2, r, 1) \,\&\, \text{Small}_p(q_1 + q_2, r, 1). \tag{7.2.7}$$

It was already noted in Section 3.6 that conditions (3.6.5)–(3.6.7) uniquely determine the numbers $h_0, \ldots, h_{b-1}$, so the resulting function $F[b]$ for prime b is singlefold.

In the representation (5.5.8) for the function NextT, one can choose for z any number that is not less than the length of the tuples with ciphers p and t. To obtain a singlefold representation, one can, for example, set $z = p + t$:

$$t' = \text{NextT}(p, t) \iff t' = A[b](p, t, p + t). \tag{7.2.8}$$

The same can be done in order to transform representation (5.5.16) for the function NextP into a singlefold representation.

In Section 5.5 we checked that, for given values of k, l, p, and t, the corresponding values of p_L, t_L, p_M, t_M, p_R, t_R, p', t' are unique. The value of l is not uniquely determined by the values of k, p, and t because for l one can choose any number satisfying inequalities (5.5.22). To get a singlefold representation, we replace (5.5.22) by, say, the equation

$$l = k + p + t + 3. \tag{7.2.9}$$

The value of k is uniquely determined from the values of p and t by the number of steps executed by the Turing machine before it halts. The values of p and t satisfying (5.5.36) are also uniquely determined by the position of the head when the machine halts. However, we cannot use condition (5.5.36) as it stands because it contains disjunctions. Nevertheless, it is not difficult to see that machine (5.4.10) can halt only in state q_2, so the role of (5.5.36) can be played by the condition

$$\exists kr\,[\mathrm{Elem}(\mathrm{AfterT}(k,p,t),b,r)=2]. \tag{7.2.10}$$

Finally, conditions (5.5.37) and (5.5.38) uniquely determine the initial configuration $\langle p,t\rangle$ from the values of the parameters $a_1, \ldots, a_n$.

Second technique. The second technique for constructing singlefold representations is a modification of the technique from Section 6.2. Without loss of generality, we can assume that the set $\mathfrak{M}$ is one-dimensional and has a representation

$$a\in\mathfrak{M} \iff \exists x_2\ldots x_m\,[M(a,x_2,\ldots,x_m)=0]. \tag{7.2.11}$$

Using a bounded universal quantifier, we can easily express the condition *tuple* $\langle a_2,\ldots,a_m\rangle$ *is the solution of the Diophantine equation* $M(a,x_2,\ldots,x_m)=0$ *with the least Cantor number*, namely,

$$\begin{aligned}&M(a,a_2,\ldots,a_m)=0\ \&\ \exists b\big[b=\mathrm{Cantor}_{m-1}(a_2,\ldots,a_m)\ \&\\ &\forall y<b\,\exists x_1,\ldots,x_m\,[y=\mathrm{Cantor}_{m-1}(x_2,\ldots,x_m)\ \&\ M^2(a,x_2,\ldots,x_m)=x_1+1]\big].\end{aligned} \tag{7.2.12}$$

To obtain a singlefold exponential Diophantine representation of the set $\mathfrak{M}$, it suffices to modify the technique of Section 6.2 in such a manner that all the unknowns arising from elimination of the bounded universal quantifier in (7.2.12) are uniquely determined by a and b.

The only conditions imposed on the choice of w were the inequalities (6.2.20). It is easy to convince oneself that the functions Cantor and Cantor_{m-1} are monotonically increasing in each of their arguments, and so for the numbers $x_2, \ldots, x_m$ in (7.2.12), the inequalities

$$x_i\le y<b \tag{7.2.13}$$

hold. This allows us to bound x_1 as well:

$$x_1<C^2(a,b), \tag{7.2.14}$$

where $C(a,b)$ is the polynomial obtained from $M(a,x_2,\ldots,x_m)$ by changing the signs of all of its negative coefficients and systematically replacing $x_2, \ldots, x_n$ by b. Thus, in eliminating the bounded universal quantifier from (7.2.12), we can impose on w the additional condition

$$w=b+C^2(a,b), \tag{7.2.15}$$

which uniquely determines w from a and b.

For the choice of the value of q there was also only one restriction, the divisibility condition (6.2.24). Simply by replacing the divisibility symbol by the equal sign, we guarantee the uniqueness of q and, by (6.2.15), the uniqueness of z_0 as well.

For the choice of $z_1, \ldots, z_m$ there was only one restriction, namely, that for $0 \le y < b$, the numbers $x_{1,y}, \ldots, x_{m,y}$ defined by (6.2.6) should satisfy the original equation (6.2.4). So by the Chinese Remainder Theorem (see the Appendix), we can impose on $z_1, \ldots, z_m$ the additional restriction

$$z_i < \binom{q}{b} \tag{7.2.16}$$

because, by (6.2.13), the right-hand side of (7.2.16) is just the product $q_0 \cdots q_{w-1}$. Let us verify that this condition makes the choice of $z_1, \ldots, z_m$ unique.

Let the prime p and the number k be chosen so that

$$p^k \mid \binom{q}{b}. \tag{7.2.17}$$

All the factors in the left-hand side of (6.2.13) are pairwise coprime; hence, there is a unique value of y such that $0 \le y < b$ and $p \mid q_y$, which implies that

$$p^k \mid q_y. \tag{7.2.18}$$

Condition (7.2.18) can play the role of condition (6.2.18), and the condition

$$p^k \mid \prod_{x=0}^{w-1} (z_i - x) \tag{7.2.19}$$

can play the role of the divisibility condition (6.2.27). Any common divisor of two different factors in the product in (7.2.19) is no greater than the absolute value of their difference, and hence it is less than w; on the other hand, by (6.2.24), $p > w$. Thus, in the product in (7.2.19), only one factor can be divisible by p, and hence it must be divisible by p^k as well. Let us define the numbers $x_{1,y}, \ldots, x_{m,y}$ by the inequalities (6.2.20) and the condition

$$p^k \mid z_i - x_{i,y}. \tag{7.2.20}$$

This last condition serves as a counterpart of condition (6.2.28), and clearly, the condition

$$p^k \mid z_0 - y \tag{7.2.21}$$

can play the role of (6.2.29). As in Section 6.2, we can deduce from this that the numbers $x_{1,y}, \ldots, x_{m,y}$ provide a solution of the initial equation (6.2.4); i.e., in the present case,

$$y = \text{Cantor}_{m-1}(x_{2,y}, \ldots, x_{m,y}), \tag{7.2.22}$$

$$M^2(a, x_{2,y}, \ldots, x_{m,y}) = x_{1,y} + 1. \tag{7.2.23}$$

We see that the values of $x_{1,y}, \ldots, x_{m,y}$ are uniquely determined by y and hence by p as well; in other words, in every solution, the remainder $\text{rem}(z_i, p^k)$ has the same value. Only condition (7.2.17) was imposed on p and k, and hence the remainder

$$\text{rem}\left(z_i, \binom{q}{b}\right) \tag{7.2.24}$$

also always has the same value, which by (7.2.16) is equal to z_i.

Thus, the values of all the unknowns q, w, z_1, ..., z_m are uniquely determined by the values of the parameters a and b. We have already checked that binomial coefficients and the factorial have singlefold representations. It remains to declare that in the transformed formula (7.2.12) the variables $a_2, \ldots, a_m$ are unknowns; we can do so because their values, and hence the value of b, are uniquely determined by a.

Third technique. The third technique for constructing singlefold representations is a modification of the technique from Section 6.3. Let us transform the Diophantine representation (7.2.11) into the representation

$$\forall y < 1 \, \exists x_1 \ldots x_m \, [D(a, y, x_1, \ldots, x_m) = 0], \tag{7.2.25}$$

where

$$D(a, y, x_1, \ldots, x_m) = M^2(a, x_2, \ldots, x_m) + (x_1 - 1)^2. \tag{7.2.26}$$

The role of the additional term in (7.2.26) will become clear later; evidently, the universal quantifier over y is fictitious since y can only have the value zero. The universal quantifier was introduced only to make it possible to use the notation introduced in Section 6.3. That is, the representation (7.2.25) is a formal counterpart of (6.3.2) with $b = 1$.

It was shown in Section 6.3 that formula (6.3.42) is equivalent to formula (6.3.35); in the present case the formula

$$\exists x_1 < w+1 \ldots \exists x_m < w+1 \, [D(a, 0, x_1, \ldots, x_m) = 0] \tag{7.2.27}$$

is equivalent to the formula

$$2^{I(w)} \mid S(w), \tag{7.2.28}$$

where $I(w)$ and $S(w)$ are defined by formulas (6.3.3), (6.3.15), (6.3.16), (6.3.17), (6.3.23), (6.3.24), (6.3.30), (6.3.34), and (6.3.36) with $b = 1$, $y = 0$.

Because of the term $(x_1 - 1)^2$ in (7.2.26), assertions (7.2.27) and (7.2.28) are false when $w = 0$. Thus, if the equation in (7.2.11) has a solution, then there is a number u such that for $w \leq u$ formulas (7.2.27) and (7.2.28) are false, but for $w > u$ they are true. Clearly, the number u is therefore uniquely determined, and formula (7.2.11) is equivalent to the formula

$$\exists u \left[2^{I(u)} \nmid S(u) \,\&\, 2^{I(u+1)} \mid S(u+1)\right]. \tag{7.2.29}$$

In Section 6.3 we checked that I and S are Diophantine functions. Moreover, $I(w)$ has an explicit representation defined by (6.3.30), (6.3.15), and (6.3.16), so that the exponential Diophantine representation for I has no unknowns at all and hence is singlefold. Furthermore, by (6.3.36), (6.3.34), and (6.3.23),

$$S(w) = C(w) = \binom{2K(w)}{K(w)}. \tag{7.2.30}$$

We have already verified that the relations of divisibility and non-divisibility and binomial coefficients have singlefold exponential Diophantine representations, and $K(w)$ has the explicit representation (6.3.49). Thus, we can consider (7.2.29) as a generalized singlefold exponential Diophantine representation of the set M.

7.3 Gaussian integer counterpart of Hilbert's Tenth Problem

In this section we wish to deal with solutions of equations in the ring of *Gaussian integers*, i.e., numbers of the form $p + qi$, where p and q are integers and i is the imaginary unit. For this reason, in this section we deviate from the convention introduced in Section 1.3 according to which lowercase italic Latin letters denote natural numbers. By $\sqrt{z}$ we shall understand a complex number w with non-negative real part such that $w^2 = z$.

So, let us consider an equation

$$D(x_1, \ldots, x_m) = 0, \tag{7.3.1}$$

where D is a polynomial with Gaussian integer coefficients, and where we are interested in the solvability of this equation in Gaussian integers $x_1, \ldots, x_m$. Clearly, equation (7.3.1) has a solution if and only if the equation

$$D(p_1 + q_1 i, \ldots, p_m + q_m i) = 0 \tag{7.3.2}$$

has a solution in rational integers $p_1, \ldots, p_m, q_1, \ldots, q_m$. In general, (7.3.2) may have Gaussian integer coefficients, but we can easily separate the real and imaginary parts. Thus, let

$$\begin{aligned} D(p_1 + q_1 i, \ldots, p_m + q_m i) = D_{\mathrm{R}}(p_1, \ldots, p_m, q_1, \ldots, q_m) \\ + D_{\mathrm{I}}(p_1, \ldots, p_m, q_1, \ldots, q_m) i \end{aligned} \tag{7.3.3}$$

where D_{R} and D_{I} are polynomials with rational integer coefficients. Thus, we have reduced the problem of the solvability of (7.3.1) in Gaussian integers to the problem of the solvability in rational integers of the following equation whose coefficients are rational integers:

$$D_{\mathrm{R}}^2(p_1, \ldots, p_m, q_1, \ldots, q_m) + D_{\mathrm{I}}^2(p_1, \ldots, p_m, q_1, \ldots, q_m) = 0. \tag{7.3.4}$$

So, a positive solution of Hilbert's Tenth Problem would also yield a method for testing the solvability of Diophantine equations in Gaussian integers. The aim of this section is to show that the negative solution of Hilbert's Tenth Problem implies a *negative solution of its counterpart for Gaussian integers.*

For this purpose, we are going to establish that *the set of rational integers is Diophantine in the ring of Gaussian integers.* The latter notion needs to be defined precisely, and we shall do so in a general setting.

Below, we shall write $\mathbb{N}$ for the set of natural numbers, and $\mathbb{Z}$, $\mathbb{G}$ for the rings of rational and Gaussian integers, respectively. Let the ring R be an extension of $\mathbb{Z}$, let C be a subring of R, let A and X be subsets of R, and let

$$D(a_1, \ldots, a_n, x_1, \ldots, x_m) = 0 \tag{7.3.5}$$

be a polynomial equation with coefficients from C and variables separated into parameters $a_1, \ldots, a_n$ and unknowns $x_1, \ldots, x_m$. Such an equation defines the set of n-tuples:

$$\begin{aligned} \{ \langle a_1, \ldots, a_n \rangle \mid a_1, \ldots, a_n \in A \\ \& \exists x_1 \ldots x_m \, [x_1, \ldots, x_m \in X \,\&\, D(a_1, \ldots, a_n, x_1, \ldots, x_m) = 0] \}. \end{aligned} \tag{7.3.6}$$

We shall call a set that can be defined in this manner (A, C, X)*-Diophantine.* In this new terminology, what we have been studying in the previous chapters is the notion of being $(\mathbb{N}, \mathbb{Z}, \mathbb{N})$-Diophantine, which, by the results of Section 1.3, is equivalent to the notion of being $(\mathbb{N}, \mathbb{Z}, \mathbb{Z})$-Diophantine.

Below we establish that the set $\mathbb{Z}$ is $(\mathbb{G}, \mathbb{Z}, \mathbb{G})$-Diophantine; i.e., we find a Diophantine equation

$$Z(a, u_1, \ldots, u_l) = 0 \tag{7.3.7}$$

that is solvable in Gaussian integers $u_1, \ldots, u_l$ when $a \in \mathbb{Z}$ and unsolvable for $a \in \mathbb{G} - \mathbb{Z}$. In a manner analogous to the transformation of the single equation (1.3.2) into the system (1.3.3), this will enable us to reduce the problem of the solvability of an arbitrary Diophantine equation

$$D(y_1, \ldots, y_m) = 0 \tag{7.3.8}$$

in rational integers $y_1, \ldots, y_m$ to the solvability of the system

$$\begin{aligned} D(y_1, \ldots, y_m) &= 0 \\ Z(y_1, u_{1,1}, \ldots, u_{1,l}) &= 0 \\ &\vdots \\ Z(y_m, u_{m,1}, \ldots, u_{m,l}) &= 0 \end{aligned} \tag{7.3.9}$$

in Gaussian integers $y_1, \ldots, y_m, u_{1,1}, \ldots, u_{m,l}$. However, the technique used in passing from (1.1.1) to (1.1.2) to compress a system into a single equation cannot be used here, because for Gaussian integers u, v, $u^2 + v^2 = 0$ does not imply that $u = v = 0$. Instead, we have to use a slightly deeper algebraic fact, namely that $\sqrt{2}$ is irrational, in order to combine a system of two equations

$$\begin{aligned} D_1(x_1, \ldots, x_k) &= 0 \\ D_2(x_1, \ldots, x_k) &= 0 \end{aligned} \tag{7.3.10}$$

in Gaussian integers $x_1, \ldots, x_k$ into an equivalent equation

$$D_1^2(x_1, \ldots, x_k) + 2D_2^2(x_1, \ldots, x_k) = 0. \tag{7.3.11}$$

Applying this transformation repeatedly, we can compress system (7.3.9) into the desired equation whose solvability in Gaussian integers is equivalent to the solvability of the original equation (7.3.8) in rational integers.

In our proof that $\mathbb{Z}$ is $(\mathbb{G}, \mathbb{Z}, \mathbb{G})$-Diophantine, we will again use the sequences $\alpha_b(n)$ introduced in (2.1.4). At this point we will only need the single sequence with $b = 4$, so we shall omit the subscript. Then the characteristic equation (2.1.12) takes the form

$$x^2 - 4xy + y^2 = 1. \tag{7.3.12}$$

By (2.1.11), this equation has solutions of the form

$$x = \alpha(n+1), \qquad y = \alpha(n) \tag{7.3.13}$$

and

$$x = \alpha(n), \qquad y = \alpha(n+1) \tag{7.3.14}$$

for $n = 0, 1, \ldots$ We wish to study the solutions of (7.3.12) in Gaussian integers. To begin with, in this context we naturally need to consider, in addition to (7.3.13) and (7.3.14), the negative solutions

$$x = -\alpha(n+1), \qquad y = -\alpha(n) \tag{7.3.15}$$

and

$$x = -\alpha(n), \qquad y = -\alpha(n+1) \tag{7.3.16}$$

for $n = 0, 1, \ldots$ We can combine (7.3.13) with (7.3.16) and (7.3.14) with (7.3.15) by defining

$$\alpha(n) = -\alpha(-n) \tag{7.3.17}$$

for $n = -1, -2, \ldots$ With this extension of the definition, the recurrent relation

$$\alpha(n+1) = 4\alpha(n) - \alpha(n-1) \tag{7.3.18}$$

is valid for all rational integers n, and equation (2.1.9) holds for all such n:

$$\mathrm{A}(n) = \begin{pmatrix} \alpha(n+1) & -\alpha(n) \\ \alpha(n) & -\alpha(n-1) \end{pmatrix} = \Xi^n. \tag{7.3.19}$$

Let us show that *all solutions of equation* (7.3.12) *in Gaussian integers x and y are exhausted by the sequence* (7.3.13) *and* (7.3.14) *with arbitrary rational integral values of n.* Let x and y satisfy (7.3.12). Let us consider the matrix

$$\mathrm{B} = \begin{pmatrix} 4x - y & -x \\ x & -y \end{pmatrix} \tag{7.3.20}$$

and show that for some rational integer m, either

$$\mathrm{B} = \mathrm{A}(m) \tag{7.3.21}$$

or

$$\mathrm{B} = -\mathrm{A}(m). \tag{7.3.22}$$

In the former case x and y satisfy (7.3.13) with $n = m$, and in the latter case they satisfy (7.3.14) with $n = -m$.

How can one find the value of m for a given x and y? The matrix Ξ has an eigenvector

$$\begin{pmatrix} 1 \\ 2-\sqrt{3} \end{pmatrix}, \tag{7.3.23}$$

corresponding to the eigenvalue

$$\lambda = 2 + \sqrt{3}. \tag{7.3.24}$$

The vector (7.3.23) is an eigenvector of the matrix B as well, and the corresponding eigenvalue is

$$\mu = \lambda x - y. \tag{7.3.25}$$

We now define the value of m by the inequalities

$$1 \leq |\mu|\lambda^{-m} < \lambda \tag{7.3.26}$$

and consider the matrix

$$\mathrm{BA}(-m) = \mathrm{B}\Xi^{-m} = \begin{pmatrix} 4u-v & -u \\ u & -v \end{pmatrix}, \tag{7.3.27}$$

where

$$\begin{aligned} u &= -\alpha(m-1)x + \alpha(m)y, \\ v &= -\alpha(m)x + \alpha(m+1)y. \end{aligned} \tag{7.3.28}$$

We have

$$u^2 - 4uv + v^2 = \det(\mathrm{B}\Xi^{-m}) = 1. \tag{7.3.29}$$

If $u = 0$, then either $v = 1$ or $v = -1$ and, correspondingly, either (7.3.22) or (7.3.21) is valid; thus, it will suffice to verify that u and v cannot have any other value.

Let $u \neq 0$, so that by (7.3.29) either

$$v = (2 + \nu)u \tag{7.3.30}$$

or

$$v = (2 - \nu)u, \tag{7.3.31}$$

where

$$\nu = \sqrt{3 + u^{-2}}. \tag{7.3.32}$$

It is easy to see that the inequality

$$\mathrm{Re}(\nu) \geq \sqrt{2} \tag{7.3.33}$$

holds and that for large $|u|$, the value of ν is close to $\sqrt{3}$:

$$|\nu - \sqrt{3}| = \left|\sqrt{3 + u^{-2}} - \sqrt{3}\right| = \frac{1}{|u|^2|\nu + \sqrt{3}|} < \frac{1}{2|u|^2}. \tag{7.3.34}$$

The vector (7.3.23) is evidently also an eigenvector of the matrix (7.3.27) and corresponds to the eigenvalue

$$\lambda u - v = \mu\lambda^{-m}, \tag{7.3.35}$$

so that by (7.3.26),

$$1 \leq |\lambda u - v| < \lambda. \tag{7.3.36}$$

It follows from (7.3.30) and (7.3.34) that

$$|\lambda u - v| = |\sqrt{3} - \nu||u| < \frac{1}{2|u|}, \tag{7.3.37}$$

which contradicts the first inequality in (7.3.36). Thus, of the two possible equations (7.3.30) and (7.3.31), only the latter could hold.

It follows from (7.3.31) and (7.3.33) that

$$|\lambda u - v| = |\sqrt{3} + \nu||u| > |\sqrt{3} + \sqrt{2}||u|, \tag{7.3.38}$$

which together with the second inequality in (7.3.36) gives the bound $|u|^2 < 2$, so that u can have only the values ± 1 or $\pm i$. In the former case, $\nu = 2$ and $v = 0$, which would violate the second inequality in (7.3.36). In the latter case, $\nu = \sqrt{2}$ and $v = \pm(2 - \sqrt{2})i$, which is not a Gaussian integer. Thus, inequality (7.3.31) cannot be true either, and the contradiction proves that $u = 0$.

So, we have checked that all solutions of equation (7.3.12) in Gaussian x and y are in fact real; also, we know that this equation has arbitrarily large solutions. How can we use these facts to obtain a definition of the rational integers?

Clearly, arithmetical operations, and in particular division, do not lead out of the set of real numbers. Therefore, in any solution of the system

$$x^2 - 4xy + y^2 = 1 \tag{7.3.39}$$

$$u^2 - 4uv + v^2 = 1 \tag{7.3.40}$$

$$v = qy \tag{7.3.41}$$

$$y \neq 0 \tag{7.3.42}$$

in Gaussian integers q, u, v, x, y, all of these numbers must be real.

Let us define the three-argument relation grem by the equivalence

$$\text{grem}(a, b, c) \iff \exists z\big[a = b + cz \ \& \ |a| < |c|\big]. \tag{7.3.43}$$

(In contrast to (1.6.4), this relation does not define a function, because the choice of a is not unique.) Let us check that if b and c are rational integers, then grem(a, b, c) implies that a is also a rational integer. In (7.3.43), let $z = g + hi$ with rational integers g and h. Then

$$c^2 > |a|^2 = |b + cz|^2 = |b + c(g + hi)|^2 = (b + cg)^2 + c^2h^2, \tag{7.3.44}$$

whence $h = 0$, and hence z and a are real.

So, in any solution of the system consisting of conditions (7.3.39)–(7.3.41) and

$$p = q + zy, \tag{7.3.45}$$

$$|p| < |y|, \tag{7.3.46}$$

all the unknowns are rational integers.

We'll see below that the system (7.3.39)–(7.3.41), (7.3.45)–(7.3.46) has a solution for any odd rational integer value for p. After that, obtaining the entire set $\mathbb{Z}$ presents no difficulty. However, condition (7.3.46) does not suit us because the straightforward way to show that it is $(\mathbb{G}, \mathbb{Z}, \mathbb{G})$-Diophantine requires us to show that $\mathbb{N}$, or at least $\mathbb{Z}$, is itself $(\mathbb{G}, \mathbb{Z}, \mathbb{G})$-Diophantine. That is why we replace condition (7.3.46) by the two conditions

$$x = 4(3p^2 + 1)w, \tag{7.3.47}$$

$$x \neq 0. \tag{7.3.48}$$

It follows from (7.3.47) and (7.3.48) that

$$|x| = 4|3p^2 + 1||w| > 6|p|, \tag{7.3.49}$$

$$|x| = 4|3p^2 + 1||w| \geq 4, \tag{7.3.50}$$

which together with (7.3.39) implies that

$$|y| = \left|2 \pm \sqrt{3 + x^{-2}}\right| |x| \geq \left(2 - \sqrt{3 + \frac{1}{16}}\right) |x| = \frac{1}{4}|x| > |p|. \tag{7.3.51}$$

We eliminated the relation $<$ from (7.3.46) at the cost of the appearance of the relation $\neq$ in (7.3.48). We replace this latter condition by the equation

$$rx + (8s + 3)t = 1, \tag{7.3.52}$$

which has no solution in Gaussian integers r, s, and t when $x = 0$.

So, we have constructed the system of Diophantine equations (7.3.39)–(7.3.41), (7.3.45), (7.3.47), and (7.3.52) and established that in any of its Gaussian integer solutions, the value of p must be real. Let us now verify that for every odd rational integer value of p, this system has a solution in q, r, s, t, u, v, x, y, z. Without loss of generality, we will assume that $p > 0$ (otherwise it would be sufficient to find a solution for $-p$ and just change the signs of q, v, x, y, and z).

Consider the sequence

$$\alpha(0), \alpha(1), \ldots, \alpha(n), \ldots \tag{7.3.53}$$

Due to the defining recurrence, this sequence is purely periodic modulo $4(3p^2 + 1)$. Since $\alpha(0) = 0$ by definition, there are infinitely many multiples of $4(3p^2 + 1)$ in the sequence (7.3.53). It is easy to deduce from the recurrent relation (or from (2.2.7)) that

$$\alpha(n) \equiv n \pmod 2. \tag{7.3.54}$$

Therefore, there is an odd value of n such that $n > 3$ and

$$\alpha(n + 1) \equiv 0 \pmod{4(3p^2 + 1)}. \tag{7.3.55}$$

Setting

$$x = \alpha(n - 1), \qquad y = \alpha(n), \tag{7.3.56}$$

equation (7.3.39) is satisfied, and we can find w satisfying (7.3.47). Choosing s such that x and $8s + 3$ are coprime, we can find r and t satisfying (7.3.52).

Setting

$$u = \alpha(pn+1), \qquad v = \alpha(pn), \tag{7.3.57}$$

equation (7.3.40) is satisfied. By (2.3.9),

$$v \equiv px^{pn-1}y \pmod{y^2}, \tag{7.3.58}$$

so we can find a value of q satisfying (7.3.41), and this value will satisfy the congruence

$$q \equiv px^{pn-1} \pmod{y}. \tag{7.3.59}$$

By (7.3.39),

$$x^2 \equiv 1 \pmod{y}, \tag{7.3.60}$$

so that it follows from (7.3.59), (7.3.60), and the fact that pn is odd that

$$q \equiv p \pmod{y}, \tag{7.3.61}$$

and we can find z satisfying (7.3.45).

To obtain a $(\mathbb{G}, \mathbb{Z}, \mathbb{G})$-Diophantine representation of the set $\mathbb{Z}$, it suffices to use the method of (7.3.11) to combine equations (7.3.39)–(7.3.41), (7.3.45), (7.3.47), (7.3.52), and the equation

$$2a + 1 = p, \tag{7.3.62}$$

and to declare that a is a parameter and that the remaining variables are unknowns.

7.4 Homogeneous equations and rational solutions

In this book we have been mainly concerned with solutions of Diophantine equations in natural numbers. Hilbert in his Tenth Problem asked about solutions in integers. However, Diophantus himself sought solutions in rational numbers. If we are given a Diophantine equation

$$D(\chi_1, \ldots, \chi_m) = 0 \tag{7.4.1}$$

whose unknowns $\chi_1, \ldots, \chi_m$ range over the rationals, then we may consider instead the equation

$$H(r_1, \ldots, r_m, q) = 0 \tag{7.4.2}$$

with integer unknowns, defined by

$$H(r_1, \dots, r_m, q) = q^k D\left(\frac{r_1}{q}, \dots, \frac{r_m}{q}\right), \tag{7.4.3}$$

where k is the degree of the polynomial D. The polynomial H is homogeneous of degree k, and hence equation (7.4.2) certainly has the *trivial solution* $r_1 = \dots = r_m = q = 0$. For this reason, when one speaks of the solvability or unsolvability of a homogeneous Diophantine equation, one has in mind the existence or absence of a non-trivial solution. The main aim of this section is to establish the equivalence of two decision problems: *the problem of determining the existence of a rational solution for arbitrary Diophantine equations* and *the problem of determining the existence of a nontrivial integer solution for homogeneous Diophantine equations.*

We start by reducing the latter of these problems to the former. Let

$$F(r_1, \dots, r_m) = 0 \tag{7.4.4}$$

be a homogeneous Diophantine equation. If it has a non-trivial solution with $r_i \neq 0$, then the equation

$$F(\chi_1, \dots, \chi_{i-1}, 1, \chi_{i+1}, \dots, \chi_m) = 0 \tag{7.4.5}$$

has a rational solution

$$\chi_j = \frac{r_j}{r_i}. \tag{7.4.6}$$

On the other hand, every solution of equation (7.4.5) in rational $\chi_1, \dots, \chi_{i-1}, \chi_{i+1}, \dots, \chi_m$ produces an integer solution of equation (7.4.4) in which

$$r_i = d, \quad r_j = d\chi_j, \qquad j = 1, \dots, i-1, i+1, \dots, m, \tag{7.4.7}$$

where d is a common denominator of $\chi_1, \dots, \chi_{i-1}, \chi_{i+1}, \dots, \chi_m$. Thus, in order to determine whether equation (7.4.4) has a non-trivial solution, it suffices to determine whether at least one of the m equations

$$\begin{gathered} F(1, \chi_2, \dots, \chi_{m-1}, \chi_m) = 0, \\ \vdots \\ F(\chi_1, \chi_2, \dots, \chi_{m-1}, 1) = 0. \end{gathered} \tag{7.4.8}$$

has a rational solution.

(Note that here we have a reduction of a somewhat different nature than those we have dealt with hitherto. Up until now, in reducing a decision problem P_1 to another decision problem P_2, we found for each individual subproblem of P_1 an equivalent subproblem of P_2. We could do the same in this case by multiplying the left-hand sides in (7.4.8), thus reducing the solvability of (7.4.4) to the solvability of a single equation in rational numbers. However, we need not do so; clearly, if there were a method for recognizing the existence of rational solutions, we could apply it m times to the equations of (7.4.8) and thus determine whether equation (7.4.4) has a non-trivial solution. Computability theory deals with reductions of even more general kinds, for example, reductions with an unbounded number of individual subproblems of problem P_2 required to be answered in order to find the answer to a single subproblem of problem P_1.)

The inverse reduction is less evident. If equation (7.4.1) has a solution in rational $\chi_1, \ldots, \chi_m$, then equation (7.4.2) has a non-trivial solution, but the converse is not true in general because in a non-trivial solution of equation (7.4.2), the value of q might be equal to 0. We need to combine equation (7.4.2) with the condition

$$r_1^2 + \cdots + r_m^2 \neq 0 \Longrightarrow q \neq 0, \tag{7.4.9}$$

and this combination must take the form of a homogeneous equation.

Here we can once again use equation (7.3.12). We know from Section 2.1 that this equation has solutions with arbitrarily large values of x. On the other hand, the equation

$$x^2 - 4xy + y^2 = 0 \tag{7.4.10}$$

has only the trivial solution, because

$$x^2 - 4xy + y^2 = (x - (2 + \sqrt{3})y)(x - (2 - \sqrt{3})y). \tag{7.4.11}$$

This implies that for arbitrary $q \neq 0$, the equation

$$x^2 - 4xy + y^2 = q^4 \tag{7.4.12}$$

has a solution with arbitrarily large x, while for $q = 0$, the only solution is $x = 0$. This allows us to rewrite condition (7.4.9) in the form of the inequality

$$x \geq r_1^2 + \cdots + r_m^2. \tag{7.4.13}$$

To obtain the required homogeneity, it suffices to apply the Four Squares Theorem (see the Appendix). Namely, we can introduce new unknowns and give the following explicit expressions for x and y:

$$x = r_1^2 + \cdots + r_m^2 + s_1^2 + s_2^2 + s_3^2 + s_4^2, \tag{7.4.14}$$

$$y = t_1^2 + t_2^2 + t_3^2 + t_4^2. \tag{7.4.15}$$

Substitution of these expressions in (7.4.12) results in the desired homogeneous equation

$$\begin{aligned}(r_1^2 + \cdots + r_m^2 + s_1^2 + s_2^2 + s_3^2 + s_4^2)^2 \\ - 4(r_1^2 + \cdots + r_m^2 + s_1^2 + s_2^2 + s_3^2 + s_4^2)(t_1^2 + t_2^2 + t_3^2 + t_4^2) \\ + (t_1^2 + t_2^2 + t_3^2 + t_4^2)^2 = q^4\end{aligned} \tag{7.4.16}$$

solvable with respect to s_1, s_2, s_3, s_4, t_1, t_2, t_3, t_4 for any q, r_1, ..., r_m, and for which $q \neq 0$ in every non-trivial solution.

It remains to combine the two equations (7.4.2) and (7.4.16). We cannot directly apply the technique of Section 1.2 because if the homogeneous equations (1.2.1) are of different degrees, then the corresponding equation (1.2.2) is not homogeneous. However, this obstacle is easy to overcome since the system of two homogeneous equations

$$\begin{aligned}D_1(x_1, \ldots, x_m) = 0 \\ D_2(x_1, \ldots, x_m) = 0\end{aligned} \tag{7.4.17}$$

of degrees k_1 and k_2 respectively is equivalent to the single homogeneous equation

$$D_1^{2k_2}(x_1, \ldots, x_m) + D_2^{2k_1}(x_1, \ldots, x_m) = 0 \tag{7.4.18}$$

of degree $2k_1k_2$.

Thus, a positive solution of the restriction of Hilbert's Tenth Problem to the case of homogeneous equations would supply us with a method for determining the existence of rational solutions for arbitrary Diophantine equations.

Exercises

1. Analogously to Card, one can consider ECard, the number of solutions of an exponential Diophantine equation. Show that the set $\text{ECard}^{-1}(\mathfrak{A})$ is semidecidable if and only if either $\mathfrak{A}$ is empty or $\mathfrak{A} = \{\eta \mid \eta \geq m, \eta \in \mathfrak{N}\}$, where m is a finite element of the set $\mathfrak{N}$. (A rigorous statement of the problem requires considering $\text{ECard}^{-1}(\mathfrak{A})$ as a set of natural numbers, for which we first need to introduce the notion of a code of an exponential Diophantine equation).

2. Let $F(\eta) = c_0 + c_1\binom{\eta}{1} + \cdots + c_m\binom{\eta}{m}$. We treat F as a mapping from $\mathfrak{N}$ into itself by defining $\binom{\infty}{k} = \infty$. Show that for any (exponential) Diophantine equation $D_1 = 0$, one can construct another (exponential) Diophantine equation $D_2 = 0$ such that $\mathrm{Card}(D_2) = F(\mathrm{Card}(D_1))$ $(\mathrm{ECard}(D_2) = F(\mathrm{ECard}(D_1)))$.

3. Show that for any $b > 1$, every Diophantine set has a singlefold exponential Diophantine representation with base b.

4. Show that the union of two disjoint sets each of which has a singlefold exponential Diophantine representation also has such a representation.

5. Show that every Diophantine set $\mathfrak{D}$ has a singlefold "almost Diophantine" representation of the form

$$a \in \mathfrak{D} \iff \exists x_1 \ldots x_m y\, [D(a, x_1, \ldots, x_m) = y + 4^y].$$

6. Let A be the ring of algebraic integers of the quadratic field $\mathbb{Q}(\sqrt{Z})$, where the integer Z is not a perfect square. Show that the set $\mathbb{N}$ is $(A, \mathbb{Z}, A)$-Diophantine.

7. The method for decreasing the degree of an equation presented in Section 1.2 leads, in general, to non-homogeneous equations. Show that every homogeneous Diophantine equation can be transformed into an equivalent homogeneous equation of degree 4.

Open questions

1. Is it true that if $\mathfrak{A}$ is a non-empty set with semidecidable $\mathrm{Card}^{-1}(\mathfrak{A})$, then $\mathfrak{A} = \{\eta \mid \eta \geq m,\ \eta \in \mathfrak{N}\}$ for some finite m?

2. Is it possible to find effectively, for an arbitrary exponential Diophantine equation $E = 0$, a Diophantine equation $D = 0$ such that $\mathrm{ECard}(E) = \mathrm{Card}(D)$?

3. Is it true that the union of any two sets, each of which has a singlefold Diophantine representation, also has such a representation?

Unsolved problems

1. Is it true that every Diophantine set has a singlefold Diophantine representation?

2. Let $\mathbb{O}$ be the ring of algebraic integers of a finite extension of the field of rational numbers $\mathbb{Q}$. Is it true that the set $\mathbb{Z}$ is $(\mathbb{O}, \mathbb{O}, \mathbb{O})$-Diophantine?

Commentary

Hilbert [1900] devoted less space to the Tenth Problem than to any of the others. Constance Reid, in her biography of Hilbert [1970], explains that the Tenth Problem was one of several listed in the published version of Hilbert's lecture but omitted from the oral presentation for lack of time. Thus, we have neither a detailed motivation for the importance of the problem nor an explanation as to why its statement is so definite.

A plausible explanation might be as follows. Judging from the spirit of the paper, Hilbert presumably expected a positive solution of the Tenth Problem. We have seen that had that been the case, we would not only have possessed a method for recognizing solvability in integers of individual equations, but also methods for recognizing solvability in natural, rational, or Gaussian integers of individual equations and systems of equations. Hilbert might have thought that in restricting himself to the case of integer solutions, he was posing the most difficult problem, which would have automatically provided solutions to other similar problems.

However, now we know that the Tenth Problem in the form stated by Hilbert is unsolvable, and this does *not* produce as a corollary the unsolvability of all related problems. That is why one can interpret Hilbert's Tenth Problem in a more general setting, including all those problems positive solutions of which would have followed straightforwardly from a positive solution of the Tenth Problem in the narrower sense.

Among such problems, we should first mention the analogs of the Tenth Problem for solutions in other rings of algebraic numbers. In proving the undecidability for the case of Gaussian integers in Section 7.3, we followed Denef [1975]. In fact, that paper contains a more general result about arbitrary quadratic extensions (see Exercise 7.6). This result was further generalized in a joint paper by Denef and Lipshitz [1978], and in that paper Unsolved Problem 7.2 was posed. Further progress in that direction has been obtained by Denef [1980], Pheidas [1988], Shapiro and Shlapentokh [1989], and Shlapentokh [1989].

As a contrasting result, we can mention Rumely [1986], which gives a method for testing for the existence of solutions of Diophantine equations in *arbitrary* algebraic integers. However, this result goes beyond the framework of even the extended interpretation of the Tenth Problem, because there is no straightforward method for transforming a Diophantine equation whose unknowns range over algebraic integers of arbitrary degree into a Diophantine equation with rational integer unknowns.

Methods for solving Diophantine equations in rational numbers would have very important applications. However, results in this area are scarce, which can likely be explained by the fact that researchers are deprived of such an important tool as the use of divisibilities and congruences. The fact that Diophantine equations with rational unknowns can be reduced to homogeneous equations has most likely been known for a long time, but it seems that a proof was not published until relatively recently. In Section 7.4 we followed the proof given by R. M. Robinson (see Smoryński [1987]). For other problems equivalent to solvability of Diophantine equations, see Kim and Roush [1989].

Lipshitz [1977, 1978a, 1978b] studied decision problems connected with systems of Diophantine conditions of special form with coefficients and unknowns from certain rings of algebraic numbers. Davis and Putnam [1963], Denef [1978a, 1978b], Pheidas [1987a, 1987b], and Shlapentokh [1990] studied decision problems connected with solving Diophantine equations in rings whose elements are not numbers.

The existence of singlefold exponential Diophantine representations was established by Matiyasevich [1974] in a manner similar to the second technique from Section 7.2. The first and third techniques have prototypes in Matiyasevich [1976] and [1979] respectively.

The undecidability of $\text{Card}^{-1}(\{1\})$ (as a consequence of the undecidability of Hilbert's Tenth Problem) was established by Adler [1969b]. The undecidability of $\text{Card}^{-1}(\mathfrak{A})$ in the general case was proved by Davis [1972]. The proof was simplified by Smoryński [1977] (see also Davis [1977]), who also gave a more detailed characterization of the corresponding sets of exponential Diophantine equations by making use of the existence in the exponential case of singlefold representations (see Exercise 7.1). Open questions 7.1, 7.2, and a number of others were posed in the same paper.

8 Diophantine Complexity

In this chapter we consider quantitative improvements of some of our previous results.

8.1 Principal definitions

So far almost all of our results have had a qualitative character, e.g., we showed that some properties and relations are Diophantine and established the undecidability of certain decision problems connected with Diophantine equations. However, most of these results can be improved by adding a quantitative aspect.

Consider the main notion studied in this book, the notion of Diophantine set. Recall that a set $\mathfrak{D}$ of n-tuples of natural numbers is called Diophantine if

$$\langle a_1, \ldots, a_n \rangle \in \mathfrak{D} \iff \exists x_1 \ldots x_m \,[D(a_1, \ldots, a_n, x_1, \ldots, x_m) = 0], \quad (8.1.1)$$

where D is a polynomial with integer coefficients. For every Diophantine set $\mathfrak{D}$, one can pose the following two questions:

(a) What is the minimum possible degree with respect to the unknowns of a polynomial D in a Diophantine representation (8.1.1) of the set $\mathfrak{D}$?
(b) What is the minimum possible value of m, the number of unknowns in a representation (8.1.1)?

The former characteristic will be called the *order* of the set $\mathfrak{D}$, denoted by $\text{ord}(\mathfrak{D})$, the latter the *rank* of $\mathfrak{D}$, denoted by $\text{rank}(\mathfrak{D})$.

Clearly, the minimal degree and the minimal number of unknowns may well be achieved with different polynomials D, so one can be interested in the *relative order* $\text{ordrank}(\mathfrak{D}, m)$, the minimum degree of the polynomial D in a representation (8.1.1) with m unknowns, and the *relative rank* $\text{rankord}(\mathfrak{D}, d)$, the minimum number of unknowns in a representation (8.1.1) with a polynomial D of degree d; for $m < \text{rank}(\mathfrak{D})$, we take $\text{ordrank}(\mathfrak{D}, m) = \infty$; similarly, $\text{rankord}(\mathfrak{D}, d) = \infty$ for $d < \text{ord}(\mathfrak{D})$.

Diophantine sets fall into disjoint classes according to the values of $\text{ord}(\mathfrak{D})$ or $\text{rank}(\mathfrak{D})$. *A priori*, one might have expected that all these classes would be non-empty, but the existence of universal Diophantine equations shows that this is not the case: *there exist constants* $\mathbf{d}$ *and* $\mathbf{m}$ *such that every Diophantine set has a representation* (8.1.1) *in which at the same time the degree of the equation with respect to the unknowns is no greater than* $\mathbf{d}$ *and the number of unknowns is no greater than* $\mathbf{m}$. It is clear that for these numbers $\mathbf{d}$ and $\mathbf{m}$, one can simply choose the degree and the number of unknowns in some particular universal equation (such equations were constructed in Sections 4.5 and 6.5). In other words, for every Diophantine set

$$\text{ordrank}(\mathfrak{D}, \mathbf{m}) \le \mathbf{d}, \tag{8.1.2}$$

$$\text{rankord}(\mathfrak{D}, \mathbf{d}) \le \mathbf{m}. \tag{8.1.3}$$

So, the traditional classification of Diophantine equations as "equations of degree 1," "equations of degree 2," ..., "equations of degree d," ... and "equations in 1 unknown," "equations in 2 unknowns," ..., "equations in m unknowns," ... makes sense only for small values of d and m.

A pair $\langle \mathbf{d}, \mathbf{m} \rangle$ that satisfies inequality (8.1.2) (or, equivalently, inequality (8.1.3)) for every Diophantine set $\mathfrak{D}$ will be called a *universal complexity bound* for Diophantine sets. Numbers $\mathbf{d}$ and $\mathbf{m}$ satisfying the inequalities

$$\text{ord}(\mathfrak{D}) \le \mathbf{d} \tag{8.1.4}$$

and

$$\text{rank}(\mathfrak{D}) \le \mathbf{m} \tag{8.1.5}$$

for every Diophantine set $\mathfrak{D}$ will be called, respectively, a *universal order bound* and a *universal rank bound* for Diophantine sets.

Our first quantitative result appeared in Section 1.5, where it was shown that the number 4 is a universal order bound. No easy technique for reducing the number of unknowns has been found so far. Of course, one could find a universal rank bound by estimating the number of unknowns in a universal equation like that constructed in Section 4.5 or 6.5. This would lead to a value of several hundred, because we did not take care to use unknowns in an "economical" manner.

We may also consider another, less conventional measure of the complexity of polynomials, and hence of Diophantine sets, namely, oper($\mathfrak{D}$), the least number of arithmetic operations required for calculating the value of the polynomial D in the Diophantine representation (8.1.1) of the set $\mathfrak{D}$. Evidently, there is no universal bound on this number for sets of arbitrary dimension, but once again, the existence of a universal Diophantine equation shows that for each particular dimension n, there is a *universal bound* $\mathbf{p}_n$ *on the number of operations required.*

As an example, let us take $\mathfrak{D}$ to be the set of all numbers having at least $10^{10}\mathbf{p}_1$ prime factors. It turns out that verification of this property for any particular element of the set $\mathfrak{D}$ requires carrying out no more than $\mathbf{p}_1$ operations!

Now that we have one additional complexity measure, we can introduce the *relative orders* ordoper, ordrankoper, *the relative ranks* rankoper, rankordoper, and *the relative number of operations* operord, operrank, operordrank.

Each individual Diophantine representation (8.1.1) generates yet another complexity measure, reflecting the rate of growth of the "least" solution x_1, x_2, ..., x_m as a function of the parameters. By "least," one can understand a solution with minimum value of $\max(x_1, \ldots, x_m)$, $x_1 + \cdots + x_m$, $\mathrm{Cantor}_m(x_1, \ldots, x_m)$, or some other similar function $F(x_1, \ldots, x_m)$ for which the number of solutions of the inequality $F(x_1, \ldots, x_m) < b$ is finite for every b. A function $T(a_1, \ldots, a_n)$ will be called a *probe* for the representation (8.1.1) if in addition to (8.1.1), the following representation holds:

$$\langle a_1, \ldots, a_n \rangle \in \mathfrak{D} \iff \exists x_1 \ldots x_m \, [D(a_1, \ldots, a_n, x_1, \ldots, x_m) = 0 \;\&\; F(x_1, \ldots, x_m) \leq T(a_1, \ldots, a_n)]. \tag{8.1.6}$$

The term *probe* reflects the fact that in order to decide whether the tuple $\langle a_1, \ldots, a_n \rangle$ belongs to the set $\mathfrak{D}$, it suffices to "probe" the finite number of tuples satisfying the inequality $F(x_1, \ldots, x_m) \leq T(a_1, \ldots, a_n)$.

It should be emphasized that in contrast to order and rank, we are only speaking of a probe for *some particular Diophantine representation of a set*, rather than for the set itself. One obstacle to applying the notion directly to sets is the lack of a linear order on the set of probe functions. We could introduce a partial order $\ll$ by saying that $T_1 \ll T_2$ if for all sufficiently large a_1, ..., a_n the inequality $T_1(a_1, \ldots, a_n) \leq T_2(a_1, \ldots, a_n)$ holds, and then call a function, say, a *minimal probe function* for a given Diophantine set $\mathfrak{D}$ if it is a probe function for some representation of $\mathfrak{D}$ and is minimal with respect to $\ll$ as compared with all other representations of $\mathfrak{D}$. The reason this approach fails is not at all obvious: *there are Diophantine sets with no minimal probe function* (see Exercise 8.11 for an even stronger result).

So far we have been considering the original definition of a Diophantine set via representations of the form (8.1.1). However, as we saw in Chapters 1–3 and 6, there are wider classes of formulas that also represent only Diophantine sets. For each such class, one can introduce corresponding complexity measures. As an example, we consider exponential Diophantine representations.

Consideration of

$$\langle a_1, \dots, a_n \rangle \in \mathfrak{D}$$
$$\iff \exists x_1 \dots x_m \left[E_1(a_1, \dots, a_n, x_1, \dots, x_m) = E_2(a_1, \dots, a_n, x_1, \dots, x_m) \right] \quad (8.1.7)$$

suggests the notion of *exponential rank* $\text{erank}(\mathfrak{D})$. Since every Diophantine representation becomes an exponential Diophantine representation once the negative terms are transposed to the other side of the equal sign, $\text{erank}(\mathfrak{D}) \leq \text{rank}(\mathfrak{D})$. In Section 8.2 we establish a *universal exponential rank bound*: $\text{erank}(\mathfrak{D}) \leq 3$.

A kind of counterpart for exponential Diophantine equations to the notion of the degree of a polynomial equation is provided by the notion of the *iteration level of the equation* in (8.1.7), defined as follows:

(a) constants are of iteration level 0;
(b) variables are of iteration level 1;
(c) the iteration level of $E_1 + E_2$, $E_1 E_2$, and $E_1 = E_2$ is equal to the maximum of the iteration levels of E_1 and E_2;
(d) the iteration level of $E_1^{E_2}$ is equal to $\max(i_1, 1) + i_2$, where i_1 and i_2 are the iteration levels of E_1 and E_2, respectively.

One might also introduce $\text{iter}(\mathfrak{D})$, the *iteration level of the set* $\mathfrak{D}$, equal to the least possible iteration level of the equation in an exponential Diophantine representation of the set $\mathfrak{D}$. However, it is easy to see that $\text{iter}(\emptyset) = \text{iter}(\mathbb{N}) = 0$, and $\text{iter}(\mathfrak{D}) = 1$ for any other Diophantine set $\mathfrak{D}$. That is why it is only the *relative iteration level* itererank and the *relative exponential rank* erankiter that are of interest.

8.2 A bound for the number of unknowns in exponential Diophantine representations

In this section we show that *every Diophantine set has an exponential Diophantine representation with only three unknowns.*

Relying on the fact that Cantor_n is not merely a Diophantine function but is actually a polynomial with rational coefficients, it suffices to give a proof for the one-dimensional case.

Let $\mathfrak{M}$ be a Diophantine set of natural numbers. As with the third technique for constructing singlefold representations from Section 7.2, we are going to use a modification of the technique from Section 6.3. For this purpose, we begin with the

Diophantine representation (7.2.11) of the set $\mathfrak{M}$. It was shown in Section 7.2 that

$$a \in \mathfrak{M} \iff \exists w \left[2^{I(w)} \mid S(w) \right], \tag{8.2.1}$$

where $I(w)$ and $S(w)$ are defined by formulas (6.3.3), (6.3.15), (6.3.16), (6.3.17), (6.3.23), (6.3.24), (6.3.30), (6.3.34), and (6.3.36) with $b = 1$, $y = 0$. In Section 6.3 we agreed not to show explicitly that $I(w)$ and $S(w)$ also depend on a. To simplify the notation even further, we will no longer exhibit the dependence on w either. Representation (8.2.1) of the set $\mathfrak{M}$ contains just one unknown, but this representation is not exponential Diophantine, because it contains the relation of divisibility $\mid$ and a binomial coefficient (7.2.30). Elimination of the relation of divisibility according to (1.6.3) requires the introduction of a new unknown, and elimination of the binomial coefficient according to (3.4.2) and (3.3.8) results in yet another three unknowns. So if we wish to obtain a representation with only three unknowns, we must be more economical. By (7.2.30),

$$S = \binom{2K}{K}. \tag{8.2.2}$$

Clearly,

$$(u+1)^{2K} = pu^K + q, \tag{8.2.3}$$

where

$$p = \sum_{i=K}^{2K} \binom{2K}{i} u^{i-K}, \tag{8.2.4}$$

$$q = \sum_{i=0}^{K-1} \binom{2K}{i} u^{i}. \tag{8.2.5}$$

If u is large enough, for example, if

$$2^{2K} \leq u, \tag{8.2.6}$$

then

$$q < u^K, \tag{8.2.7}$$

because

$$u^K \geq 2^{2K} u^{K-1} = \sum_{i=0}^{2K} \binom{2K}{i} u^{K-1}. \tag{8.2.8}$$

It is easy to see that the values of p and q are uniquely determined by conditions (8.2.3) and (8.2.7).

Clearly,

$$p \equiv \binom{2K}{K} \pmod{u}, \tag{8.2.9}$$

so if

$$2^I \mid u, \tag{8.2.10}$$

then the divisibility condition in (8.2.1) is equivalent to

$$2^I \mid p. \tag{8.2.11}$$

Thus, we do not need to use an unknown explicitly set equal to the binomial coefficient. This leads to the following representation for the set $\mathfrak{M}$.

$$a \in \mathfrak{M} \iff \exists pquw \left[(u+1)^{2K} = pu^K + q \,\&\, 2^{2K} \leq u \,\&\, q < u^K \,\&\, 2^I \mid u \,\&\, 2^I \mid p\right]. \tag{8.2.12}$$

Introducing one additional unknown, we can replace condition (8.2.11) by the equation

$$p = 2^I r. \tag{8.2.13}$$

Instead of (8.2.6), we use the stronger inequality

$$2^{2K+I+2} \leq u \tag{8.2.14}$$

from which, similarly to (8.2.8), we obtain the inequality

$$2^{I+2} q < u^K, \tag{8.2.15}$$

which is stronger than inequality (8.2.7). By (8.2.15), (8.2.3), and (8.2.14),

$$2^{I+2} q < u^K \leq \frac{(u+1)^{2K}}{u^K} - 1 \leq p \leq \frac{(u+1)^{2K}}{u^K} < 2u^K, \tag{8.2.16}$$

and by (8.2.13),

$$q < r, \tag{8.2.17}$$

$$r < u^K. \tag{8.2.18}$$

Thus, if inequality (8.2.14) is valid, then (8.2.4), (8.2.5), and (8.2.13) imply (8.2.17), and (8.2.17) implies (8.2.7). This leads to the new representation for $\mathfrak{M}$:

$$a \in \mathfrak{M} \iff \exists pqruw\,[(u+1)^{2K} = pu^K + q \,\&\, 2^{2K+I+2} < u \\ \&\, 2^I \mid u \,\&\, p = 2^I r \,\&\, q < r]. \qquad (8.2.19)$$

Now we can easily obtain a representation with three unknowns by replacing inequality (8.2.17) by the equation

$$r = q + s + 1, \qquad (8.2.20)$$

using equations (8.2.13) and (8.2.20) to eliminate p and r, and using the equation

$$u = 2^{2K+I+2} \qquad (8.2.21)$$

to eliminate u. The resulting representation

$$a \in \mathfrak{M} \iff \exists rsw\,[(2^{2K+I+2}+1)^{2K} = 2^I(q+s+1)2^{(2K+I+2)K} + q] \qquad (8.2.22)$$

is still not quite the desired exponential Diophantine representation because, recalling (6.3.49),

$$K = \frac{K_1 - K_2}{K_3 - K_4}, \qquad (8.2.23)$$

and (as defined in Section 2.5) an exponential Diophantine equation is not permitted to include subtractions or divisions.

In order to eliminate the divisions, we define u by

$$u = 2^{(K_3-K_4)T}, \qquad (8.2.24)$$

where

$$T = K_1 + I + 2, \qquad (8.2.25)$$

instead of using (8.2.21). Then, instead of the equation in (8.2.22), we get the equation

$$\left(2^{(K_3-K_4)T}+1\right)^{\frac{2K_1-2K_2}{K_3-K_4}} = 2^I(q+s+1)2^{(K_1-K_2)T} + q. \qquad (8.2.26)$$

Raising both sides to the power $K_3 - K_4$, we obtain the equation

$$\left(2^{(K_3-K_4)T}+1\right)^{2K_1-2K_2} = \left(2^I(q+s+1)2^{(K_1-K_2)T} + q\right)^{K_3-K_4}. \qquad (8.2.27)$$

To eliminate subtractions, we first multiply both sides by

$$\left(2^{(K_3-K_4)T}+1\right)^{2K_2}\left(2^I(q+s+1)2^{(K_1-K_2)T}+q\right)^{K_4} \tag{8.2.28}$$

and thus obtain the equation

$$\begin{aligned}\left(2^{(K_3-K_4)T}+1\right)^{2K_1}&\left(2^I(q+s+1)2^{(K_1-K_2)T}+q\right)^{K_4}\\ &=\left(2^{(K_3-K_4)T}+1\right)^{2K_2}\left(2^I(q+s+1)2^{(K_1-K_2)T}+q\right)^{K_3}.\end{aligned} \tag{8.2.29}$$

Finally, we multiply both sides by

$$2^{(2(K_1+K_2)K_4+K_2K_3)T} \tag{8.2.30}$$

and obtain the desired exponential Diophantine equation

$$\begin{aligned}2^{K_2(K_3+K_4)T}&\left(2^{K_3T}+2^{K_4T}\right)^{2K_1}\left(2^I(q+s+1)2^{K_1T}+q2^{K_2T}\right)^{K_4}\\ &=2^{2K_1K_4T}\left(2^{K_3T}+2^{K_4T}\right)^{2K_2}\left(2^I(q+s+1)2^{K_1T}+q2^{K_2T}\right)^{K_3}.\end{aligned} \tag{8.2.31}$$

This equation has a solution in q, s, and w if and only if a belongs to the set $\mathfrak{M}$.

Exercises

1. Show that the 5-place relation $(a+1) \mid (b-c)$ & $d > e$ has a Diophantine representation with only 1 unknown.

2. Let $S_1, \ldots, S_k$ be polynomials with integer coefficients. Show that the relation

$$\exists x_1 \ldots x_k \left[\mathop{\&}_{i=1}^{k} S_i(a_1,\ldots,a_n)=x_i^2\right]$$

has a Diophantine representation with only 1 unknown.

3. Verify that if p is a prime, $l = \deg_p(q) \equiv 1 \pmod 2$, and $k > 0$, then the single divisibility condition

$$p^{2kl} \mid a^2q+b^2$$

implies the two divisibility conditions

$$p \mid a, \qquad p^{kl} \mid b.$$

Generalize this property to obtain a single divisibility condition implying m divisibility conditions.

4. Use the previous exercise to replace the m divisibility conditions (6.2.26) by a single divisibility condition.

5. Show that every Diophantine set can be represented in Davis normal form

$$a \in \mathfrak{D} \iff \exists w \, \forall z \le w \, \exists yx \, [D(a, w, x, y, z) = 0]$$

with only two variables in the scope of the second existential quantifier.

6. Show that every Diophantine set has a representation of the form

$$a \in \mathfrak{D} \iff \exists xy \, \forall z \le B(a, x, y) \, [D(a, x, y, z) > 0],$$

where B and D are polynomials with integer coefficients.

7. Show that every Diophantine set has a representation of the form

$$a \in \mathfrak{D} \iff \exists x \, \forall y \le B(a, x) \, \forall z \le C(a, x, y) \, [D(a, x, y, z) > 0],$$

where B, C, and D are polynomials with integer coefficients.

8. Show that every Diophantine set has generalized Diophantine representations of the forms

$$a \in \mathfrak{D} \iff \exists x \mathop{\&}_{i=1}^{k} \exists yz \, [D_i(a, x, y, z) > 0]$$

and

$$a \in \mathfrak{D} \iff \exists xy \mathop{\&}_{i=1}^{k} \exists z \, [D_i(a, x, y, z) > 0].$$

9. By Exercise 4.2, there is a number m such that every Diophantine set can be represented as the set of all natural numbers assumed by some polynomial with integer coefficients in m variables. Show that for every m, there is a set that can be represented as the set of all values of a polynomial whose coefficients are *natural numbers*, but such that in any such representation of that set, the number of variables cannot be less than m.

10. Improve the main result of Section 8.2 by showing that every Diophantine set has a singlefold unary exponential Diophantine representation with three unknowns.

11. Show that for every Diophantine function $T(a,b)$ defined for all values of a and b, one can find a one-dimensional Diophantine set $\mathfrak{D}$ with Diophantine complement such that for every Diophantine representation

$$D(a, x_1, \ldots, x_m) = 0 \qquad (*)$$

of (the complement of) the set $\mathfrak{D}$, the following is impossible: there is no other Diophantine representation

$$E(a, y_1, \ldots, y_n) = 0 \qquad (**)$$

of (the complement of) the set $\mathfrak{D}$ and number a_0 such that for every a, x_1, ..., x_m satisfying $(*)$ and the inequality $a > a_0$, there are numbers $y_1, \ldots, y_n$ satisfying $(**)$ and the inequality

$$T(a, y_1 + \cdots + y_n) < x_1 + \cdots + x_m.$$

Unsolved problem

1. Let **D** denote the class of all relations $\mathcal{R}$ having representations of the form

$$\mathcal{R}(a_1, \ldots, a_n) \iff \exists x_1 \ldots x_m \, [D(a_1, \ldots, a_n, x_1, \ldots, x_m) = 0$$
$$\& \, x_1 + \cdots + x_m \leq 2^{|a_1 + \cdots + a_n|^k}],$$

where D is a polynomial with integer coefficients and $|a|$ denotes the length of the binary representation of the number a. Is it true that $\mathbf{D} = \mathbf{NP}$, where **NP** is the class of relations recognizable by a nondeterministic Turing machine in a number of steps bounded by a polynomial in the length of the input?

(A rigorous definition of the class **NP** can be found, for example, in Garey and Johnson [1979]. Note that the definition of the class **D** does not presuppose that

$$\mathcal{R}(a_1, \ldots, a_n) \iff \exists x_1 \ldots x_m \, [D(a_1, \ldots, a_n, x_1, \ldots, x_m) = 0].)$$

Commentary

Quantitative aspects of Diophantine and related representations have been of interest since the very beginning of investigations aimed at the negative solution of Hilbert's Tenth Problem. Raphael M. Robinson [1956] showed that one can always take $m = 4$ in the Davis normal form soon after it was introduced (see the Commentary to Chapter 5). He later [1972] improved this result to $m = 3$, and in Matiyasevich [1972b] it was shown that one can take $m = 2$ (see Exercise 8.5). Other arithmetical representations with a small number of quantifiers (both universal and existential) were proposed by Matiyasevich and Robinson [1974], Jones [1981], and Jones, Levitz, and Wilkie [1986].

After it was established that all semidecidable sets are Diophantine, investigators began seeking to determine the Diophantine rank of particular sets as well as the least possible universal rank bound. This last is naturally connected with the construction of universal Diophantine equations, and, as was explained in the Commentary to Chapter 4, the best to date (1992) universal rank bound of 9 unknowns was obtained by Matiyasevich. A detailed proof was given in Jones [1982b]. That same paper also gives the following universal complexity bounds for Diophantine sets:

$$\langle 4,58\rangle,\ \langle 8,38\rangle,\ \langle 12,32\rangle,\ \langle 16,29\rangle,$$
$$\langle 20,28\rangle,\ \langle 24,26\rangle,\ \langle 28,25\rangle,\ \langle 36,24\rangle,$$
$$\langle 96,21\rangle,\ \langle 2668,19\rangle,\ \langle 2\times 10^5,14\rangle,$$
$$\langle 6.6\times 10^{43},13\rangle,\ \langle 1.3\times 10^{44},12\rangle,$$
$$\langle 4.6\times 10^{44},11\rangle,\ \langle 8.6\times 10^{44},10\rangle,$$
$$\langle 1.6\times 10^{45},9\rangle.$$

Jones also obtained the universal bound $\mathbf{p}_1 = 100$ for the number of operations.

As for the ranks of particular sets, it was the set of all primes that excited the most interest, as was quite natural. For a considerable period, the special character of the primes made it possible to obtain a better estimate for the rank of the set of primes than for the universal rank bound. At present (1992), the least known number of variables in a Diophantine representation of the set of primes is actually the same as the best available universal rank bound, namely 9, but of course the construction and proof for the primes are simpler than in the general case (see the Commentary to Chapter 3).

Some ideas concerning methods for reducing the number of unknowns can be obtained from Exercises 8.1–8.8.

The bound on universal exponential rank presented in Section 8.2 was obtained by Matiyasevich [1979]. It was established in that paper that there are even singlefold exponential Diophantine representations with three unknowns. This result was further improved by Jones and Matiyasevich [1982b] to unary singlefold representations (see Exercise 8.10).

There are few general results about Diophantine sets with small (exponential) rank. It is evident that Diophantine sets of rank 1 are decidable. It was shown by Levitz [1985] that sets of unary (base 2) exponential rank 1 are also decidable. The structure of sets definable by arithmetical formulas with a single quantifier (existential or universal) was described by R. M. Robinson [1951]. These results were further developed by Tung [1988].

As was stated in the Commentary to Chapter 1, the universal bound $\mathbf{d} = 4$ for the order of Diophantine sets was obtained long before the beginning of systematic research on the class of Diophantine sets. Moreover, as was also explained there, this result is within 1 of being best possible. The question of whether $\mathbf{d} = 3$ is a universal order bound was stated as Unsolved Problem 1.1.

Unsolved Problem 8.1 was posed by Adleman and Manders [1976] (see also Manders and Adleman [1978]) and by Venkatesan and Rajagopalan [1992]. They introduced the notion of a *nondeterministic Diophantine machine.* Each such machine is defined by a parametric Diophantine equation

$$D(a_1, \ldots, a_n, x_1, \ldots, x_m) = 0 \tag{8.2.1}$$

and behaves as follows: for input data represented by the values of the parameters $a_1, \ldots, a_n$, the machine "guesses" values of $x_1, \ldots, x_m$ (that is why it is nondeterministic) and returns "YES" or "NO" according to whether (1) does or does not hold. The technique of Chapter 5 allows one to show easily that nondeterministic Diophantine machine are as powerful as nondeterministic Turing machines, i.e., machines that may have several instructions with equal left-hand sides and may guess which of them to use at each step. Unsolved Problem 8.1 is about comparison of the "speed" of nondeterministic Diophantine and Turing machines. A positive solution of Problem 8.1 would provide a number-theoretic characterization of the class **NP**, in which there is currently a great deal of interest. Partial progress in this direction is reported by Adleman and Manders [1976] (see also Manders and Adleman [1978]). The class **NP** was characterized via arithmetical representations with bounded universal quantifiers by Kent and Hodgson [1982], Yukna [1982, 1983], and Hodgson and Kent [1983].

It was shown by Vinogradov and Kosovskiĭ [1975] that one can obtain the well-known Grzegorczyk hierarchy by imposing bounds on the size of the unknowns in Diophantine representations.

9 Decision Problems in Calculus

In this chapter we consider several decision problems connected with the continuum of real numbers $\mathbb{R}$. Lower-case italic Latin letters will continue to denote natural numbers; for real numbers we will use lower-case Greek letters.

9.1 Diophantine real numbers

Decision problems are unusual in conventional treatments of calculus. Most likely this is explained by the fact that the principal object of study in calculus is the continuum of real numbers $\mathbb{R}$. To specify a real number according to any of the usual definitions requires an infinite amount of information, while every individual subproblem of a decision problem must be specified by a finite amount of information, as was explained in Section 1.1. The finiteness of the input information is required by the fact that we want to get the answer after a finite number of elementary steps. For example, it is useless to pose the following problem: at the start, the infinite tape of a Turing machine is filled with the decimal digits of a real number; the machine is to determine in a finite number of steps whether the real number is rational.

Thus, if we want individual subproblems to be specified by real parameters, we must exploit some method for specifying real numbers by a finite amount of information. Such a method was proposed by Turing and is, naturally, connected with Turing machines. Namely, a real number μ can be specified by a Turing machine that operates for an infinitely long period of time, successively writing more and more digits of the number μ. The Turing machine itself is specified, as usual, by a finite set of instructions. So, it is possible to pose the problem of deciding *from a given set of instructions* for a machine whether or not it specifies a rational real number.

Without leaving the language of Diophantine equations, we can define a *Diophantine real number* as a Diophantine equation with four parameters

$$D(a, b, c, d, x_1, \dots, x_m) = 0 \tag{9.1.1}$$

having the following two properties:

(a) *for every natural number d, there are natural numbers a_d, b_d, c_d such that the equation*

$$D(a_d, b_d, c_d, d, x_1, \dots, x_m) = 0 \tag{9.1.2}$$

has a solution in x_1, ..., x_m;

(b) *if*

$$D(a', b', c', d', x'_1, \ldots, x'_m) = 0 \tag{9.1.3}$$

and

$$D(a'', b'', c'', d'', x''_1, \ldots, x''_m) = 0, \tag{9.1.4}$$

then

$$\frac{a' - b'}{c' + 1} - \frac{a'' - b''}{c'' + 1} \leq \frac{1}{d' + 1} + \frac{1}{d'' + 1}. \tag{9.1.5}$$

Taking consecutively $d = 0, 1, \ldots,$ we can find natural numbers a_0, b_0, c_0, a_1, b_1, c_1, ... for which equation (9.1.2) has solutions and consider the sequence of rational numbers

$$\frac{a_0 - b_0}{c_0 + 1}, \frac{a_1 - b_1}{c_1 + 1}, \ldots \tag{9.1.6}$$

By (9.1.5), this sequence satisfies the Cauchy convergence criterion and hence converges to some real number α. Also by (9.1.5), this number doesn't depend on the particular choice of a_d, b_d, and c_d, and so we can say that α is the real number specified by equation (9.1.1).

It is not difficult to establish that all real algebraic numbers can be specified as Diophantine real numbers (see Exercise 9.1). Using the technique of Chapter 3 for working with tuples of arbitrary length and the technique from Chapter 6 for eliminating bounded universal quantifiers, one can construct Diophantine real numbers corresponding to classical mathematical constants such as π, the base of natural logarithms, and so on (see Exercises 9.2–9.4).

It is not difficult to introduce arithmetic operations on the set of Diophantine real numbers. For example, it is easy to see that the sum of two Diophantine real numbers

$$D_1(a, b, c, d, x_1, \ldots, x_m) = 0 \tag{9.1.7}$$

and

$$D_2(a, b, c, d, x_1, \ldots, x_m) = 0 \tag{9.1.8}$$

can be defined as the Diophantine real number

$$D_1^2(p_1,q_1,r_1,2d+1,x_1,\dots,x_m)+D_2^2(p_2,q_2,r_2,2d+1,y_1,\dots,y_m)$$
$$+\big((r_2+1)p_1+(r_1+1)p_2-a\big)^2+\big((r_2+1)q_1+(r_1+1)q_2-b\big)^2$$
$$+\big((r_1+1)(r_2+1)-c\big)^2=0. \quad (9.1.9)$$

(We treat equation (9.1.9) as an equation with parameters a, b, c and unknowns p_1, p_2, q_1, q_2, r_1, r_2, x_1, $\dots$, x_m, y_1, $\dots$, y_m.) If we fix a universal Diophantine polynomial, then we can enumerate the set of all Diophantine real numbers. It can be verified that under such an enumeration, arithmetic operations on the set of Diophantine real numbers have Diophantine counterparts on the set of the codes of Diophantine real numbers. For example, there is a Diophantine function that transforms codes of equations (9.1.7) and (9.1.8) into a code of equation (9.1.9). It is possible to go further in this direction and find Diophantine functions corresponding to square root, logarithm, sine, and other elementary functions.

So, we are now in a position to pose decision problems, individual subproblems of which use real numbers in their specifications. It turns out that even the simplest are undecidable.

For example, the property *is a Diophantine real number* is undecidable because the equation

$$(a+b)D(x_1,\dots,x_m)=0 \quad (9.1.10)$$

is a Diophantine real number if and only if the equation

$$D(x_1,\dots,x_m)=0 \quad (9.1.11)$$

has no solution. We can overcome this difficulty by taking the position that algorithms for decision problems connected with real numbers are permitted to provide any answer, or even no answer, when the input numbers are not codes of Diophantine real numbers.

Even with such freedom of behavior on "incorrect" initial data, the problem of equality of a Diophantine real number to zero turns out to be undecidable. In fact, let us be given some equation (9.1.11). The formula

$$a=1\;\&\;b=0$$
$$\&\;\forall y<c\;\exists x_1\dots x_m\,[y=\mathrm{Cantor}_m(x_1,\dots,x_m)\;\&\;D(x_1,\dots,x_m)\neq 0]$$
$$\&\;\big[c=d\vee\exists x_1,\dots,x_m\,[c=\mathrm{Cantor}_m(x_1,\dots,x_m)\;\&\;D(x_1,\dots,x_m)=0]\big]$$

can be transformed into an equivalent Diophantine equation that is easily verified to be a Diophantine real number that is equal to 0 if and only if equation (9.1.11) has no solutions.

Thus, we see that while certain real numbers can be specified by a finite amount of information in the form of Diophantine real numbers, real numbers remain complicated objects from an algorithmic viewpoint. Many decision problems in calculus turn out to be unsolvable simply due to the algorithmic complexity of the real numbers specifying individual problems. However, there are decision problems that are connected with properties of real numbers but that do not use individual real numbers in specifying individual problems. It is precisely problems of this kind that will be considered in the following sections.

9.2 Equations, inequalities, and identities in real variables

It is natural to begin with the counterpart of Hilbert's Tenth Problem for real unknowns, i.e., by considering equations of the form

$$D(\chi_1, \dots, \chi_m) = 0, \tag{9.2.1}$$

where D is a polynomial with integer coefficients and $\chi_1, \dots, \chi_m$ are real unknowns. In contrast to the case of Diophantine equations, it is possible to determine whether (9.2.1) has a real solution or not. For $m = 1$ this can be done by the well-known *Sturm method*; a far-reaching generalization of this method found by Tarski [1951] enables one to work with any number of unknowns.

Thus, if we want to establish undecidability of equations in real unknowns, we have to allow the use of some functions other than addition, subtraction, and multiplication. A simple way to achieve undecidability is to allow the use of the sine function. *Let $\mathcal{F}_0$ denote the class of functions in several variables that can be constructed by composition from the constant* 1, *addition, subtraction, multiplication, and the* sin *function. There is no method for deciding for an arbitrary given function Φ from the class $\mathcal{F}_0$ whether the equation*

$$\Phi(\chi_1, \dots, \chi_k) = 0 \tag{9.2.2}$$

has a real solution.

To prove this, we take an arbitrary Diophantine equation

$$D(x_1, \dots, x_m) = 0 \tag{9.2.3}$$

and transform it into the following equation of the form (9.2.2):

$$D^2(\chi_1^2,\dots,\chi_m^2)+\sin^2(3+\psi^2)+(7\psi^2+\omega^2-1)^2$$
$$+\sin^2\bigl((3+\psi^2)\chi_1^2\bigr)+\cdots+\sin^2\bigl((3+\psi^2)\chi_m^2\bigr)=0. \quad (9.2.4)$$

It is easy to see that every solution of equation (9.2.3) in natural numbers $x_1, \dots, x_m$ yields a solution of equation (9.2.4) in real ω, ψ, χ_1, ..., χ_m. Namely, it suffices to put

$$\psi=\sqrt{\pi-3}, \qquad \omega=\sqrt{22-7\pi}, \qquad \chi_i=\sqrt{x_i}. \quad (9.2.5)$$

On the other hand, in every solution of (9.2.4) in real ω, ψ, χ_1, ..., χ_m, it is always the case that $3+\psi^2=\pi$, and hence χ_1^2, ..., χ_m^2 are natural numbers satisfying (9.2.3). So, the undecidability of Hilbert's Tenth Problem implies the undecidability of equations of the form (9.2.2) in real numbers.

To obtain the undecidability of Diophantine equations, we had to consider equations with sufficiently many unknowns. Passing from (9.2.3) to (9.2.4) required two additional unknowns. However, we can improve the above result by replacing $\mathcal{F}_0$ by its proper subset $\mathcal{F}_1$ consisting of functions of one variable.

In order to accomplish this, it would have been very helpful to have had available a real counterpart of the function Cantor_k introduced in Section 3.1 or, more precisely, counterparts of the inverse functions $\text{Elem}_{k,1}$, ..., $\text{Elem}_{k,k}$. Indeed, suppose that we had functions E_1, ..., E_k from $\mathcal{F}_1$ such that for any real numbers χ_1, ..., χ_k, there was a real number η such that

$$\text{E}_1(\eta)=\chi_1,\ \dots,\ \text{E}_k(\eta)=\chi_k. \quad (9.2.6)$$

Then the problem of the solvability of equation (9.2.2) would have been reduced to the problem of the solvability of the equation

$$\Phi(\text{E}_1(\eta),\dots,\text{E}_k(\eta))=0 \quad (9.2.7)$$

in one unknown η.

Unfortunately, the class $\mathcal{F}_1$ does not contain functions with the properties postulated for E_1, ..., E_k. Nevertheless, we can use functions belonging to $\mathcal{F}_1$ to map $\mathbb{R}$ onto an everywhere dense subset of $\mathbb{R}^k$, which will suffice for our purpose.

Let us consider the functions

$$\mathbf{E}(\eta)=\eta\sin(\eta), \quad (9.2.8)$$
$$\mathbf{H}(\eta)=\eta\sin(\eta^3) \quad (9.2.9)$$

and verify that *for every real χ, ψ, and positive ϵ, there is an η such that*

$$|\mathbf{E}(\eta) - \chi| < \epsilon, \tag{9.2.10}$$

$$\mathbf{H}(\eta) = \psi. \tag{9.2.11}$$

We will seek such an η in the interval

$$\left[2k\pi - \frac{\pi}{2}, 2k\pi + \frac{\pi}{2}\right], \tag{9.2.12}$$

where k is a large number. When η ranges over the interval (9.2.12), $\sin(\eta)$ assumes all values between -1 and 1, and $\mathbf{E}(\eta)$ assumes all values between $-2k\pi + \pi/2$ and $2k\pi + \pi/2$. Hence, if

$$k > |\chi| + 1, \tag{9.2.13}$$

then there is a value of η_0 in the interval (9.2.12) such that

$$\mathbf{E}(\eta_0) = \chi. \tag{9.2.14}$$

Let us find a positive δ such that for every η in the interval

$$[\eta_0 - \delta, \eta_0 + \delta], \tag{9.2.15}$$

the required inequality (9.2.10) holds. Clearly, for such η,

$$|\mathbf{E}(\eta) - \chi| = |\mathbf{E}(\eta) - \mathbf{E}(\eta_0)| \leq \mathbf{E}'(\xi)\delta = \big(\sin(\xi) + \xi\cos(\xi)\big)\delta, \tag{9.2.16}$$

where ξ is also in the interval (9.2.15), and hence

$$|\mathbf{E}(\eta) - \chi| \leq (1 + \eta_0 + \delta)\delta. \tag{9.2.17}$$

Without loss of generality, we will assume that $\epsilon < 1$ and put

$$\delta = \frac{\epsilon}{2k\pi + 3}. \tag{9.2.18}$$

As η ranges over the interval (9.2.15), η^3 ranges over the corresponding interval

$$[(\eta_0 - \delta)^3, (\eta_0 + \delta)^3], \tag{9.2.19}$$

the width of which can be easily estimated from below:

$$(\eta_0 + \delta)^3 - (\eta_0 - \delta)^3 = 6\eta_0^2\delta + 2\delta^3 > \frac{6(2k\pi - \pi/2)^2}{2k\pi + 3}. \tag{9.2.20}$$

Thus, for sufficiently large k, the width of interval (9.2.19) is greater than 2π. Hence, as η ranges over the interval (9.2.15), $\sin(\eta^3)$ assumes all values between -1

and 1, and therefore the image of $\mathbf{H}$ on this same interval must certainly contain the interval $[-\eta_0 + \delta, \eta_0 - \delta]$. Thus, for sufficiently large k, the interval (9.2.15) contains an η satisfying (9.2.11) and, by the choice of δ, also satisfying (9.2.10).

We are now in a position to introduce functions that can be used to play something like the role of the functions $\mathrm{Elem}_{k,n}$. Namely, we define:

$$\mathbf{E}_1(\eta) = \mathbf{E}(\eta), \qquad \mathbf{E}_{i+1}(\eta) = \mathbf{E}_i(\mathbf{H}(\eta)). \tag{9.2.21}$$

We proceed to show by induction on k that *given real numbers* $\chi_1, \ldots, \chi_k$ *and positive* ϵ, *there is a number* η_k *such that*

$$|\mathbf{E}_i(\eta_k) - \chi_i| < \epsilon, \qquad i = 1, \ldots, k. \tag{9.2.22}$$

For $k = 1$ this follows at once from (9.2.10). Given numbers $\chi_1, \ldots, \chi_{k+1}$, we apply the induction hypothesis to find η_k such that

$$|\mathbf{E}_1(\eta_k) - \chi_2| < \epsilon, \ldots, |\mathbf{E}_k(\eta_k) - \chi_{k+1}| < \epsilon$$

and then use (9.2.10) and (9.2.11) to obtain η_{k+1} such that

$$|\mathbf{E}(\eta_{k+1}) - \chi_1| < \epsilon, \qquad \mathbf{H}(\eta_{k+1}) = \eta_k.$$

Since

$$\mathbf{E}_{i+1}(\eta_{k+1}) = \mathbf{E}_i(\mathbf{H}(\eta_{k+1})) = \mathbf{E}_i(\eta_k) \qquad i = 1, \ldots, k,$$

this gives the desired result.

Now, we cannot quite use the functions $\mathbf{E}_1, \ldots, \mathbf{E}_k$ as direct counterparts of the functions $\mathrm{E}_1, \ldots, \mathrm{E}_k$ in passing from an arbitrary equation (9.2.2) to (9.2.7). Rather, we must take advantage of specific properties of equation (9.2.3). To begin with, we replace (9.2.3) by the inequality

$$\begin{aligned} M^2(\chi_1^2, \ldots, \chi_m^2)\Big[D^2(\chi_1^2, \ldots, \chi_m^2) + \sin^2(3 + \psi^2) + (7\psi^2 + \omega^2 - 1)^2 \\ + \sin^2\big((3 + \psi^2)\chi_1^2\big) + \cdots + \sin^2\big((3 + \psi^2)\chi_m^2\big)\Big] < 1, \end{aligned} \tag{9.2.23}$$

where M is a polynomial that is never equal to zero; the particular choice of M will be described below. Clearly, if $x_1, \ldots, x_m$ satisfy equation (9.2.3) and the numbers $\chi_1, \ldots, \chi_m$, ψ, and ω are defined by (9.2.5), then inequality (9.2.23) holds. Our aim is to choose the polynomial M in such a way that for every solution of inequality (9.2.23), the numbers

$$x_i = \lceil \chi_i^2 \rfloor \tag{9.2.24}$$

will satisfy equation (9.2.3). (By $\lceil \alpha \rfloor$ we understand an integer nearest to α.)

Let inequality (9.2.23) hold. To simplify notation, we put $\mu = M(\chi_1^2, \dots, \chi_m^2)$ and $\phi = 3 + \psi^2$. The polynomial M will be chosen in such a way that

$$\mu > 6. \tag{9.2.25}$$

It then follows from (9.2.23) and (9.2.25) that

$$7\psi^2 + \omega^2 - 1 < 1/6 \tag{9.2.26}$$

and hence that ϕ lies in the interval

$$\left[3, 3\tfrac{1}{6}\right]. \tag{9.2.27}$$

Now, from (9.2.23) we get that

$$\mu^{-1} > |\sin(\phi)| = |\sin(\phi) - \sin(\pi)| = |\cos(\xi)(\phi - \pi)|, \tag{9.2.28}$$

where ξ also belongs to (9.2.27), and hence $|\cos(\xi)| > 0.5$ and

$$|\phi - \pi| < 2/\mu. \tag{9.2.29}$$

Also, from (9.2.23) we get that

$$|\sin(\phi\chi_i^2)| < 1/6 \tag{9.2.30}$$

and hence that $\phi\chi_i^2$ lies in an interval of the form

$$\left[\pi\left(k_i - \frac{1}{6}\right), \pi\left(k_i + \frac{1}{6}\right)\right]; \tag{9.2.31}$$

moreover,

$$|\phi\chi_i^2 - \pi k_i| < \frac{1}{\mu}. \tag{9.2.32}$$

From (9.2.29) and (9.2.32) we get that

$$|\chi_i^2 - k_i| = \left|\frac{\pi - \phi}{\pi}\chi_i^2 + \frac{\phi\chi_i^2 - \pi k_i}{\pi}\right| < \frac{\chi_i^2 + 1}{\mu}. \tag{9.2.33}$$

Thus, $k_i = x_i$ and we can estimate the left-hand side:

$$\begin{aligned}|D(x_1,\ldots,x_m)| &\le |D(\chi_1^2,\ldots,\chi_m^2)| + |D(x_1,\ldots,x_m) - D(\chi_1^2,\ldots,\chi_m^2)| \\ &\le \frac{1}{\mu} + \sum_{i=1}^{m}\left|\frac{\partial D}{\partial x_i}\right| |\chi_i^2 - x_i| \\ &\le \frac{1}{\mu}\left(1 + \sum_{i=1}^{m}\left|\frac{\partial D}{\partial x_i}\right| (\chi_i^2 + 1)\right).\end{aligned} \tag{9.2.34}$$

D is a polynomial, and hence so are all of its derivatives, and we can choose the polynomial M to be so large that (9.2.25) holds, and then (9.2.34) will imply that

$$|D(x_1,\ldots,x_m)| < 1 \tag{9.2.35}$$

and hence that equation (9.2.3) holds.

Now we replace χ_i by $\mathbf{E}_i(\eta)$, ψ by $\mathbf{E}_{m+1}(\eta)$, and ω by $\mathbf{E}_{m+2}(\eta)$ in (9.2.23). We obtain an inequality of the form

$$\Psi(\eta) < 1, \tag{9.2.36}$$

where $\Psi \in \mathcal{F}_1$. It is not difficult to see that this inequality holds for some η if and only if inequality (9.2.23) holds for some $\chi_1, \ldots, \chi_m, \psi$, and ω, i.e., if and only if equation (9.2.3) has a solution. So, we have established that *there is no method for determining for an arbitrary function Φ in the class $\mathcal{F}_1$ whether there exists a real number η satisfying the inequality*

$$\Phi(\eta) < 0. \tag{9.2.37}$$

To get an equation (rather than an inequality), we note that if (9.2.3) has a solution, then $\Psi(\eta)$ can be made arbitrarily close to 0 by choosing η so that $\mathbf{E}_1^2(\eta), \ldots, \mathbf{E}_m^2(\eta), \mathbf{E}_{m+1}^2(\eta)$, and $\mathbf{E}_{m+2}^2(\eta)$ are sufficiently close to $x_1, \ldots, x_m$, $\pi - 3$, and $22 - 7\pi$, respectively. On the other hand, $\Psi_1(\eta)$ can always be made arbitrarily large—for example, by choosing η so that $\mathbf{E}_{m+2}(\eta)$ is large. Thus, equation (9.2.3) has a solution if and only if the equation

$$2\Psi(\eta) = 1 \tag{9.2.38}$$

has one. This proves that *there is no method for determining for an arbitrary given function Φ in the class $\mathcal{F}_1$ whether the equation*

$$\Phi(\eta) = 0 \tag{9.2.39}$$

has a solution.

Let us extend the class of admissible functions. *Let $\mathcal{F}_2$ denote the class of all functions of one variable that can be constructed using composition from the constant* 1, *addition, subtraction, multiplication, and the functions* sin *and* abs (*absolute value*). *Then there is no method for determining for an arbitrary given function Φ in the class $\mathcal{F}_2$ whether the equation*

$$\Phi(\eta) = 0 \tag{9.2.40}$$

is an identity. To see this, it suffices to note that the equation

$$1 - \Psi(\eta) + |1 - \Psi(\eta)| = 0 \tag{9.2.41}$$

holds for all η if and only if for all η the inequality (9.2.36) *fails* to hold.

9.3 Systems of ordinary differential equations

As was stated in the previous section, the word-for-word counterpart of Hilbert's Tenth Problem for real unknowns, i.e., the problem of deciding the solvability of polynomial equations with integer coefficients in real unknowns, is decidable. To obtain undecidability in the previous section, the class of admissible equations was expanded by explicitly using the sine function. In the present section, we shall deal only with polynomial equations, but they will be ordinary differential equations or, more precisely, systems of such equations. Moreover, our unknowns will be real differentiable functions of one independent variable τ rather than real numbers; to make matters definite, we will suppose that τ ranges over the interval $[0, 1]$.

To simulate real unknowns, we will use constant functions, i.e., functions satisfying the differential equation

$$\Xi'(\tau) = 0. \tag{9.3.1}$$

It is easy to check that the unique solution of the system

$$\Pi'(\tau) = 0, \qquad \Xi'(\tau) + \Pi^2(\tau)\Xi(\tau) = 0 \tag{9.3.2}$$

with boundary conditions

$$0 \le \Pi(0) \le 4, \qquad \Xi(0) = 0, \qquad \Xi(1) = 0, \qquad \Xi'(0) = 1 \tag{9.3.3}$$

is $\Pi(\tau) = \pi$, identically, and $\Xi(\tau) = \sin(\pi\tau)/\pi$. Similarly, in every solution of the system

$$\Xi'(\tau) = 0, \qquad \Psi''(\tau) + \Pi^2(\tau)\Xi^2(\tau)\Psi(\tau) = 0 \tag{9.3.4}$$

with boundary conditions

$$0 \le \Xi(0), \qquad \Psi(0) = \Psi(1) = 0, \qquad \Psi'(0) = 1, \tag{9.3.5}$$

$\Xi(\tau)$ is identically equal to some natural number x, and $\Psi(\tau) = \sin(\pi x \tau)/\pi x$.

Thus, a Diophantine equation

$$D(x_1, \dots, x_m) = 0 \tag{9.3.6}$$

has a solution in natural numbers $x_1, \dots, x_m$ if and only if the system of differential equations

$$\begin{gathered}
\Pi'(\tau) = 0, \\
\Xi'(\tau) + \Pi^2(\tau)\Xi(\tau) = 0, \\
\Xi_1'(\tau) = 0, \\
\Psi_1''(\tau) + \Pi^2(\tau)\Xi^2(\tau)\Psi_1(\tau) = 0, \\
\vdots \\
\Xi_m'(\tau) = 0, \qquad \Psi_m''(\tau) + \Pi^2(\tau)\Xi^2(\tau)\Psi_m(\tau) = 0, \\
D\big(\Xi_1(\tau), \dots, \Xi_m(\tau)\big) = 0
\end{gathered} \tag{9.3.7}$$

has a solution on $[0,1]$ satisfying the boundary conditions

$$\begin{gathered}
0 \le \Pi(0) \le 4, \\
\Xi(0) = \Xi(1) = 0, \qquad \Xi'(0) = 1, \\
\Psi_1(0) = \Psi_1(1) = 0, \qquad \Psi_1'(0) = 1, \\
\vdots \\
\Psi_m(0) = \Psi_m(1) = 0, \qquad \Psi_m'(0) = 1, \\
\Xi_1(0) \ge 0, \dots, \Xi_m(0) \ge 0.
\end{gathered} \tag{9.3.8}$$

By introducing new unknown functions, we can easily replace these boundary conditions by additional equations. Namely, a condition of the form

$$\Phi(0) \ge 0 \tag{9.3.9}$$

can be replaced by

$$\Phi(\tau) = \mathrm{Z}^2(\tau), \tag{9.3.10}$$

because conditions of the form (9.3.9) were imposed only on functions satisfying the equation $\Phi'(\tau) = 0$; conditions of the form

$$\Phi(\alpha) = \beta \tag{9.3.11}$$

with constants α and β can be replaced by the equation

$$\Phi(\alpha) - \beta = (\tau - \alpha)\mathrm{Z}(\tau). \tag{9.3.12}$$

An equation containing Φ'' can be replaced by a system of two first-order equations in Φ' and an additional function Z in a familiar manner. So, the problem of the existence of solutions of a Diophantine equation in natural numbers can be reduced to the problem of the existence of a solution of a system of polynomial differential equations of first order. Hence, *there is no method for determining for an arbitrary system of differential equations of the form*

$$\begin{gathered} P_1\big(\tau, \Xi_1(\tau), \dots, \Xi_k(\tau), \Xi_1'(\tau)\big) = 0 \\ \vdots \\ P_k\big(\tau, \Xi_1(\tau), \dots, \Xi_k(\tau), \Xi_k'(\tau)\big) = 0, \end{gathered} \tag{9.3.13}$$

where $P_1, \dots, P_k$ *are polynomials with integer coefficients, whether the system has a solution on the interval* $[0,1]$.

It is easy to verify that if the system (9.3.13) is constructed in the manner described above, beginning with an equation (9.3.6) that does have a solution, then (9.3.13) has a solution in which not all of the functions $\Xi_1, \dots, \Xi_k$ are identically zero. Hence, the system

$$\begin{gathered} Q_1\big(\tau, \Xi_1(\tau), \dots, \Xi_k(\tau), \Xi_1'(\tau)\big) = 0 \\ \vdots \\ Q_k\big(\tau, \Xi_1(\tau), \dots, \Xi_k(\tau), \Xi_k'(\tau)\big) = 0, \end{gathered} \tag{9.3.14}$$

where

$$Q_l(\tau, \chi_1, \dots, \chi_k, \eta) = (\chi_1^2 + \dots + \chi_k^2)P_l(\tau, \chi_1, \dots, \chi_k, \eta), \quad 1 \le l \le k, \tag{9.3.15}$$

has the unique solution $\Xi_1(\tau) = \dots = \Xi_k(\tau) = 0$ if and only if the original Diophantine equation (9.3.6) has no solution. Thus, *there is no method for deciding whether an arbitrary system of differential equations of the form* (9.3.13) *has exactly one solution on the interval* $[0,1]$.

9.4 Integrability

In this section we establish the undecidability of two decision problems, one of which involves definite integrals, the other, indefinite integrals.

In Section 9.2 we described a method for constructing for an arbitrary Diophantine equation (9.2.3), a function Ψ from the class $\mathcal{F}_1$ such that:

(a) if equation (9.2.3) has a solution, then for some η_0,

$$\Psi(\eta_0) = 0.5; \tag{9.4.1}$$

(b) if equation (9.2.3) has no solution, then for all η,

$$\Psi(\eta) \geq 1. \tag{9.4.2}$$

Consider the integral

$$\int_{-\infty}^{+\infty} \frac{d\eta}{(\eta^2+1)(2\Psi(\eta)-1)^2}. \tag{9.4.3}$$

The properties of the function Ψ described above imply that this integral converges if and only if equation (9.2.3) has no solution. The function under the integral sign in (9.4.3) does not belong to the class $\mathcal{F}_1$, so we have to extend our class of functions. *Let $\mathcal{F}_3$ denote the class of functions of one real variable that can be constructed using composition from the constant* 1, *addition, subtraction, multiplication, division, and the* sin *function. Then there is no method for deciding for an arbitrary function Φ in the class $\mathcal{F}_3$ whether*

$$\int_{-\infty}^{+\infty} \Phi(\eta)\, d\eta \tag{9.4.4}$$

converges.

The next problem, which concerns indefinite integrals, will involve two classes $\mathcal{F}_4$ and $\mathcal{F}_5$ of functions of one argument. The problem in question will have the following form: *Let Φ be a function in the class $\mathcal{F}_4$. Is there a function* I *in the class $\mathcal{F}_5$ such that for all η, $\Phi(\eta) = \mathrm{I}'(\eta)$?* Of course, there must be at least one function in $\mathcal{F}_4$ with no antiderivative in $\mathcal{F}_5$, or else the answer will trivially always be positive. We will need a somewhat stronger property than this for our proof. Namely, we assume that the *class $\mathcal{F}_4$ contains a function* H *that is defined for all values of its argument, but such that for no non-empty open interval (a, b) does the class $\mathcal{F}_5$ contain a function* I *such that* $\mathrm{H}(\eta) = \mathrm{I}'(\eta)$ *for* $\eta \in (a, b)$. One candidate for such a function is e^{x^2}, because its antiderivative is not representable as the composition of elementary functions on any interval.

We will also assume that *the class $\mathcal{F}_4$ contains the class $\mathcal{F}_2$ and is closed under multiplication.* In particular then, the class $\mathcal{F}_4$ will contain the function

$$\mathrm{M}(\eta) = \begin{cases} 2 & \text{if } \eta \le \frac{3}{4}, \\ 8 - 8\eta & \text{if } \frac{3}{4} \le \eta \le 1, \\ 0 & \text{if } 1 \le \eta, \end{cases} \tag{9.4.5}$$

because

$$\mathrm{M}(\eta) = 1 + |4\eta - 4| - |4\eta - 3|. \tag{9.4.6}$$

Consider the function

$$\mathrm{T}(\eta) = \mathrm{M}\big(\Psi(\eta)\big)\mathrm{H}(\eta). \tag{9.4.7}$$

If equation (9.2.3) has no solution, then by (9.4.2) and (9.4.5), $T(\eta) = 0$ for every η and hence any constant is an antiderivative of T. On the other hand, if equation (9.2.3) has a solution, then by (9.4.1) and (9.4.5), $T(\eta) = 2\mathrm{H}(\eta)$ in some neighborhood of η_0, and hence T has no antiderivative. Thus, *given the above assumptions about the classes $\mathcal{F}_4$ and $\mathcal{F}_5$, there is no method for determining for an arbitrary function in $\mathcal{F}_4$ whether it has an antiderivative in $\mathcal{F}_5$.*

Exercises

1. Show that all real algebraic numbers are representable as Diophantine real numbers.

2. Show that if A, B, C, and E are given Diophantine functions defined for all natural numbers and if for all n,

$$|S(E(n))| < \frac{1}{n},$$

where

$$S(m) = \sum_{k=m}^{\infty} \frac{A(k) - B(k)}{C(k) + 1},$$

then $S(0)$ is representable as a Diophantine real number.

3. Show that the numbers e and π are representable as Diophantine real numbers.

4. Show how to carry the results of Section 9.3 over to the case of a single partial differential equation.

Commentary

Turing originally [1936] defined a computable real number as a number whose positional notation can be generated by a Turing machine. This notion turned out to be not quite successful, because there is no method for obtaining a Turing machine to compute the sum of two computable real numbers (in this sense) from the two given machines.

Turing soon [1937] removed this shortcoming by passing to a system with a fractional base and not requiring a one-to-one correspondence between real numbers and their positional notations. The notion of Diophantine real number introduced in Section 9.1 is equivalent to this modified notion of computable number.

For a mathematician with a conventional mathematical background, the countable set of Diophantine real numbers appears to be too meager. However, it is sufficiently rich to support the development of a meaningful theory (see, for example, Martin-Löf [1970]).

The content of Sections 9.2–9.4 is based on Richardson [1968], Adler [1969a], Caviness [1970], and Wang [1974]. The first three of these papers were written before Hilbert's Tenth Problem was shown to be undecidable. For that reason, the first two of these papers used a weaker result, namely the undecidability of exponential Diophantine equations, and therefore exponentiation was used in their constructions. The results in the third paper were conditional; i.e., they were obtained under the assumption of the undecidability of Hilbert's Tenth Problem.

The undecidability of a number of other decision problems in calculus was established on the base of the undecidability of (exponential) Diophantine equations by Denef and Lipshitz [1984, 1989], Singer [1978], Pappas [1985], Penzin [1978], and Scarpellini [1963].

The undecidability of numerous decision problems connected with real numbers places significant restrictions on the power of the systems of computer algebra that have recently become a working tool for mathematicians. These difficulties are discussed by Barton and Fitch [1972a, 1972b], McGettrick [1977], Moses [1971], and Fitch [1973].

10 Other Applications of Diophantine Representations

The diverse results presented in this Chapter have one unifying theme; namely, they can all be proved relatively easily by using the fact that every semidecidable set is Diophantine.

10.1 Diophantine games

Consider a Diophantine equation

$$D(a_1, \ldots, a_m, b_1, \ldots, b_m) = 0 \tag{10.1.1}$$

in which the variables are split into two parts; without loss of generality, we may assume that both parts contain the same number of variables, although some of them may occur vacuously. Such an equation defines a *Diophantine game* between two persons A and B, with the following rules:

on the first move, A chooses a value for a_1;

on the second move, B chooses a value for b_1;

$\vdots$

on the $(2m-1)$*st move, A chooses a value for* a_m;

on the $2m$*th move, B chooses a value for* b_m;

if the values of $a_1, \ldots, a_m, b_1, \ldots, b_m$ *thus chosen satisfy* (10.1.1), *then B is the winner; otherwise A is.*

Each Diophantine game is a finite deterministic competitive game with complete information, so one of the players has a winning strategy. This means that one player's victory is "guaranteed" independent of the moves of the other. It is easy to see that B has a winning strategy if and only if the following statement is true:

$$\forall a_1 \, \exists b_1 \, \forall a_2 \, \exists b_2 \ldots \forall a_m \, \exists b_m \, [D(a_1, \ldots, a_m, b_1, \ldots, b_m) = 0]. \tag{10.1.2}$$

For example, for the game defined by the equation

$$(b_1 + a_2)^2 + 1 = (b_2 + 2)(b_3 + 2), \tag{10.1.3}$$

player A has a winning stategy if and only if there are infinitely many primes of the form $k^2 + 1$. In spite of the fact that this problem has attracted the attention of number theorists for a long time, we still (1992) do not know which of the two players, A or B, has a winning strategy in the game defined by such a simple equation as (10.1.3).

It is easy to see that the problem of determining, given an equation (10.1.1), which of the players has a winning strategy is undecidable, even if we consider only degenerate games with vacuously occurring variables $a_1, \ldots, a_m$. So, one can pose some complexity questions; for example, how small a value for m suffices for the problem to be unsolvable?

Algorithmic difficulties may also exist for games for which we know which player has the winning strategy. Consider first the case when A has a winning strategy. To actually win, A needs to first make a "correct" choice of a_1. Then, after learning which value has been chosen for b_1, A must make a "correct" choice of a_2, and so on. Let us take for D some polynomial defining a Diophantine set $\mathfrak{D}$ of natural numbers with non-Diophantine complement:

$$a_1 \in \mathfrak{D} \iff \exists x_1 \ldots x_m [D(a_1, \ldots, a_m, x_1, \ldots, x_m) = 0]. \tag{10.1.4}$$

(The variables $a_2, \ldots, a_m$ are vacuous; examples of such sets were constructed in Sections 4.6 and 6.6.) Clearly, the set $\mathfrak{A}_1$ of "correct" choices of a_1 is the complement of $\mathfrak{D}$ and thus is not Diophantine, or, equivalently, is not semidecidable. Thus, player A has no algorithm for deciding for arbitrary a_1 whether this number is a suitable choice for the first move.

This in itself is not very interesting, because in fact player A doesn't need to be able to determine whether an arbitrary a_1 belongs to $\mathfrak{A}_1$. It would be sufficient for A to somehow be able to find one single member of $\mathfrak{A}_1$, which would then provide a first move that could be used each time the game is played.

Let us now consider the more interesting situation that occurs when B has a winning strategy. To win, player B needs, given any a_1, to find a "correct" second move b_1. Let $\mathfrak{L}_1$ be the set of all pairs $\langle a_1, b_1 \rangle$ such that b_1 is a "correct" reply to A's move a_1. Once again, player B doesn't really need to be able to decide whether an arbitrary pair $\langle a_1, b_1 \rangle$ belongs to $\mathfrak{L}_1$. It would be quite sufficient for B to be able to decide this for a subset $\mathfrak{L}_1^\star$ of the set $\mathfrak{L}_1$, just so long as for each a_1, the subset contains at least one pair $\langle a_1, b_1 \rangle$ from $\mathfrak{L}_1$. If $\mathfrak{L}_1^\star$ has this property, we say that $\mathfrak{L}_1^\star$ *traces* the set $\mathfrak{L}_1$. If B has access to a decidable set $\mathfrak{L}_1^\star$ tracing $\mathfrak{L}_1$, then player B can find a "correct" reply to any initial move a_1 by simply scanning all of the pairs $\langle a_1, 0 \rangle$, $\langle a_1, 1 \rangle, \ldots$ and checking whether each of them belongs to $\mathfrak{L}_1^\star$; eventually a pair $\langle a_1, b_1 \rangle$ from $\mathfrak{L}_1^\star$ will be encountered, and this will yield a "correct" move b_1.

Moreover, it would be enough for the set $\mathfrak{L}_1^\star$ to be semidecidable rather than decidable. In fact, let

$$B_1(a_1, b_1, x_2, \ldots, x_n) = 0 \tag{10.1.5}$$

be a Diophantine equation defining such a semidecidable set $\mathfrak{L}_1^\star$. Player B may begin scanning the numbers 0, 1, $\ldots$, treating each number as the Cantor number of an

n-tuple $\langle b_1, x_2, \ldots, x_n \rangle$ and checking whether the tuple satisfies equation (10.1.5). Clearly, the first element of an n-tuple satisfying (10.1.5) can be used for a "correct" second move.

Let us show that there is a Diophantine game with a winning strategy for player B for which there is no semidecidable set $\mathfrak{L}_1^\star$ tracing the set $\mathfrak{L}_1$ of "correct" replies. We begin with a Diophantine representation

$$a \in \mathfrak{S} \iff \exists x_2 \ldots x_m \, [P(a, x_2, \ldots, x_m) = 0] \tag{10.1.6}$$

of the simple set $\mathfrak{S}$ constructed in Section 6.6. Consider the Diophantine game defined by the equation

$$P^2(a_1 + b_1, a_2, \ldots, a_m) - b_m - 1 = 0. \tag{10.1.7}$$

It is easy to see that b_1 is a "correct" reply to a given move a_1 if and only if $a_1 + b_1 \notin \mathfrak{S}$, because then and only then will player B be able to use (10.1.6) to choose a number b_m satisfying (10.1.7) independently of the choices of $a_2, \ldots, a_m$ made by player A. Such a b_1 always exists because the set $\overline{\mathfrak{S}}$ is infinite. Thus, in fact it is player B who has a winning strategy.

Suppose now that game (10.1.7) has a tracing set $\mathfrak{L}_1^\star$. Consider the Diophantine set $\mathfrak{D}$ defined by

$$a \in \mathfrak{D} \iff \exists a_1 b_1 \, [a = a_1 + b_1 \; \& \; \langle a_1, b_1 \rangle \in \mathfrak{L}_1^\star]. \tag{10.1.8}$$

This set is a subset of $\overline{\mathfrak{S}}$ because every pair $\langle a_1, b_1 \rangle \in \mathfrak{L}_1^\star$ produces a "correct" reply. On the other hand, the set $\mathfrak{D}$ is infinite because for every a_1 there is a suitable b_1, and this contradicts our choice of $\mathfrak{S}$ as a simple set.

This contradiction shows that no tracing set can be Diophantine, i.e., semidecidable, and hence certainly not decidable. On the other hand, if player B were able to find effectively, given a_1, a "correct" $b_1 = B_1(a_1)$, then the set of pairs $\langle a_1, B_1(a_1) \rangle$ would be decidable (in the intuitive sense) and hence Diophantine. Thus, by Church's Thesis we see that *although player B has a winning reply to every first move by player A, there is no effective procedure by means of which player B can find such a reply.*

10.2 Generalized knights on a multidimensional chessboard

The squares of the conventional chessboard can be numbered by pairs $\langle q, r\rangle$ where

$$0 \leq q, \qquad 0 \leq r, \tag{10.2.1}$$

$$q \leq 7, \qquad r \leq 7, \tag{10.2.2}$$

and the rules governing knight moves can be described as follows: from a square $\langle q, r\rangle$ a knight can move to any of the 8 squares

$$\langle q \pm 2, r \pm 1\rangle, \langle q \pm 1, r \pm 2\rangle \tag{10.2.3}$$

provided that the coordinates of the new square satisfy inequalities (10.2.1)–(10.2.2).

In this section we wish to study a kind of generalized chess. The board is to be n-dimensional, and its "squares" will be designated by arbitrary n-tuples of natural numbers $\langle r_1, \dots, r_n\rangle$. So, we will have an analog of restriction (10.2.1), but restriction (10.2.2) will have no counterpart, which is appropriate because in the case of a finite board, all reasonable decision problems are decidable. The board will be populated by (generalized) knights. The rules for moving a knight $\mathfrak{K}$ are defined by a finite set

$$\langle \delta_{1,1}, \dots, \delta_{1,n}\rangle, \dots, \langle \delta_{k,1}, \dots, \delta_{k,n}\rangle \tag{10.2.4}$$

of n-tuples of natural numbers. From square $\langle r_1, \dots, r_n\rangle$ the knight $\mathfrak{K}$ can move to square $\langle r_1 + \delta_{l,1}, \dots, r_n + \delta_{l,n}\rangle$, provided, of course, that the elements of this n-tuple are non-negative.

The rules (10.2.4) define a reflexive and transitive *relation of reachability* of one square from another on the set of all squares. In contrast to the conventional knight (10.2.3), this relation need not be symmetric, so in this respect generalized knights are more like pawns.

We shall write $\mathfrak{K}(r_1, \dots, r_n)$ to denote the *reachability set*, i.e., the set of all squares that knight $\mathfrak{K}$ can visit starting from square $\langle r_1, \dots, r_n\rangle$. It is easy to see that the relation of reachability has the following *property of additivity: if*

$$\langle r', \dots, r'_n\rangle \in \mathfrak{K}(r_1, \dots, r_n), \tag{10.2.5}$$

then

$$\langle r'_1 + t_1, \dots, r'_n + t_n\rangle \in \mathfrak{K}(r_1 + t_1, \dots, r_n + t_n). \tag{10.2.6}$$

One can think of the set $\mathfrak{K}(r_1, \dots, r_n)$ as a measure of the "chess power" of the knight $\mathfrak{K}$ occupying the square $\langle r_1, \dots, r_n\rangle$. Correspondingly, one can pose various decision problems involving the comparison of the "chess power" of knights. We will deal with the following two problems:

(a) *Comparison for inclusion*: given knights $\mathfrak{K}'$ and $\mathfrak{K}''$ and a square $\langle r_1, \ldots, r_n \rangle$, determine whether the knight $\mathfrak{K}'$ "is at least as strong" as the knight $\mathfrak{K}''$, i.e.,

$$\mathfrak{K}'(r_1, \ldots, r_n) \supseteq \mathfrak{K}''(r_1, \ldots, r_n). \tag{10.2.7}$$

(b) *Comparison for equality*: given knights $\mathfrak{K}'''$ and $\mathfrak{K}''''$ and a square $\langle p_1, \ldots, p_n \rangle$, determine whether both knights have equal "chess power," i.e.,

$$\mathfrak{K}'''(p_1, \ldots, p_n) = \mathfrak{K}''''(p_1, \ldots, p_n). \tag{10.2.8}$$

It is clear that the latter problem can be reduced to the former. Namely, to check (10.2.8) it is sufficient to check (10.2.7) both with $\mathfrak{K}' = \mathfrak{K}'''$ and $\mathfrak{K}'' = \mathfrak{K}''''$ and with $\mathfrak{K}' = \mathfrak{K}''''$ and $\mathfrak{K}'' = \mathfrak{K}'''$. So, it would be sufficient to establish the undecidability of the problem of comparison for equality. However, for technical reasons we begin by establishing the undecidability of the problem of comparison for inclusion.

(It was noted above that it is essential that we do not impose a counterpart to restriction (10.2.2), i.e., that we permit the knights to move on an infinite board. It is less evident that for undecidability it is also essential that we do have a counterpart to restriction (10.2.1), i.e., that the board is bounded in one direction in every dimension. Indeed, on the board $\mathbb{Z}^n$ all moves commute, so the problem of the reachability of a square $\langle r_1'', \ldots, r_n'' \rangle$ from square $\langle r_1', \ldots, r_n' \rangle$ can be reduced to the problem of the solvability in integers of the following system of linear Diophantine equations

$$\begin{aligned} \delta_{1,1}x_1 + \cdots + \delta_{k,1}x_k &= r_1'' - r_1' \\ &\vdots \\ \delta_{1,n}x_1 + \cdots + \delta_{k,n}x_k &= r_n'' - r_n', \end{aligned} \tag{10.2.9}$$

and the set $\mathfrak{K}(r_1, \ldots, r_n)$ has a quite regular structure.)

In establishing the undecidability of the comparison of "chess power," we shall only need to consider knights defined by rules of the form (10.2.4) in which $\delta_{l,m}$ can assume only one of the three values -1, 0, 1. With this restriction, the rules for knight moves can be presented in a simple and perspicuous form: namely, the rule corresponding to an n-tuple $\langle \delta_1, \ldots, \delta_n \rangle$ will also be written as

$$R_{i_1} \ldots R_{i_a} \Longrightarrow R_{j_1} \ldots R_{j_b}, \tag{10.2.10}$$

where $i-1, \ldots, i_a, j_1, \ldots, j_b$ are indices such that $\delta_{i_1} = \cdots = \delta_{i_a} = -1$, $\delta_{j_1} = \cdots = \delta_{j_b} = 1$, and the remaining δ's are 0.

Clearly, all the information in (10.2.10) is contained in the subscripts, so the letter R adds nothing. In the future we will use other letters as well as R. However, their only purpose will be to facilitate reading. Thus, for example, the rule

$$T_{i_1} \ldots T_{i_a} \Longrightarrow S_{j_1} \ldots S_{j_b} \tag{10.2.11}$$

will be entirely equivalent to rule (10.2.10).

We will use knight moves on the board to simulate the operation of a (nondeterministic) *chess machine*. Like a Turing machine, a chess machine can be in one of finitely many *states* $Q_{i_1}, \ldots, Q_{i_l}$ (the numbering of the states need not be consecutive). However, unlike a Turing machine, the memory of a chess machine consists of finitely many *registers* $R_{j_1}, \ldots, R_{j_k}$ (the numbering of the registers need not be consecutive either; however, the numbers of the registers must be different from those of the states). Each of the registers is capable of storing an arbitrary natural number. The *instructions* of a chess machine are of the form

$$Q_{i_0} R_{i_1} \ldots R_{i_k} \Longrightarrow Q_{j_0} R_{j_1} \ldots R_{j_l}, \tag{10.2.12}$$

where all the subscripts $i_0, \ldots, i_k, j_0, \ldots, j_l$ are distinct. Instruction (10.2.12) can be performed when the machine is in state Q_{i_0} and none of the registers $R_{i_1}, \ldots, R_{i_k}$ is empty. Performing this instruction consists in decreasing the values in registers $R_{i_1}, \ldots, R_{i_k}$ by 1 and increasing the values in registers $R_{j_1}, \ldots, R_{j_l}$ by 1; afterwards the machine goes into state Q_{j_0}. An essential difference between Turing machines and chess machines is that the latter are nondeterministic; i.e., it is possible that, at a given instant, there may be more than one instruction that could be executed.

The numbers in the registers will be simulated by coordinates of squares, and a machine being in state Q_i will be simulated by the ith coordinate being equal to 1 while all of the other coordinates corresponding to states are zero. There is a natural correspondance between the operation of the machine and the moves of a knight whose rules, written in our revised notation, are identical with the machine instructions. Using this relationship, we shall freely interlace the two terminologies, i.e., that of knight moves and that of a chess machine.

Our first goal is to show that chess machines can calculate the values of certain polynomials. First of all, we need to make precise what is to be understood by a computation by such a nondeterministic device as a chess machine. Let $F(x_1, \ldots, x_m)$ be a natural number valued function of natural number arguments, and let $\mathfrak{K}$ be a chess machine for which there are designated an initial state Q_b, a final state Q_e, input registers I_{i_1}, ..., I_{i_m}, and an output register O_k. To calculate $F(x_1, \ldots, x_m)$, one places the knight on square

$$\langle r_1, \ldots, r_n \rangle \tag{10.2.13}$$

with $r_b = 1$, $r_{i_l} = x_l$, $l = 1, \ldots, m$, and all other coordinates equal to 0. We say that a chess machine computes the function F if the following three conditions hold:

(a) *if the square* $\langle r'_1, \ldots, r'_n \rangle$ *is reachable from* (10.2.13), *then* $r'_k \leq F(x_1, \ldots, x_n)$;
(b) *for every* y *such that* $y \leq F(x_1, \ldots, x_n)$, *there is a square* $\langle r'_1, \ldots, r'_n \rangle$ *reachable from square* (10.2.13) *such that* $r'_k = y$, $r'_e = 1$;
(c) *the state* Q_e *is not present in the left-hand sides of the instructions.*

It is easy to see that because of the additivity property, a chess machine can only compute functions that are monotonically nondecreasing for every argument.

The following machine with initial state Q_1, input registers I_4, I_5, output register O_6, and final state Q_3 computes the sum:

$$\begin{aligned} Q_1 I_4 &\Longrightarrow Q_2 O_6 \\ Q_1 I_5 &\Longrightarrow Q_2 O_6 \\ Q_1 &\Longrightarrow Q_3 \\ Q_2 &\Longrightarrow Q_1. \end{aligned} \tag{10.2.14}$$

A chess machine that computes the product is a bit more complicated:

$$\begin{aligned} Q_1 I_6 &\Longrightarrow Q_2 \\ Q_2 I_7 &\Longrightarrow Q_3 R_8 O_9 \\ Q_3 &\Longrightarrow Q_2 \\ Q_3 &\Longrightarrow Q_4 \\ Q_4 R_8 &\Longrightarrow Q_1 I_7 \\ Q_1 &\Longrightarrow Q_4 \\ Q_1 &\Longrightarrow Q_5. \end{aligned} \tag{10.2.15}$$

Here Q_1 is the initial state, I_6 and I_7 are input registers, O_9 is the output register, and Q_5 is final state. Let x and y be the initial values in input registers I_6 and I_7, and let u and v be numbers such that $u \leq x$ and $v \leq y$. Suppose that the machine passed through the following sequence of states:

$$Q_1\big((Q_2Q_3)^y(Q_4Q_1)^y\big)^u(Q_2Q_3)^v(Q_4Q_1)^vQ_5. \tag{10.2.16}$$

Then the register O_9 would contain the number $uy + v$. Thus, at the moment that the machine enters the final state Q_5, the register O_9 can contain any number not greater than xy. On the other hand, the value in the register O_9 can increase by 1 only when the machine passes from state Q_2 into Q_3; however, for every sequence of states

$$(Q_2Q_3)^v, \tag{10.2.17}$$

we have the inequality $v \leq y$, because at any moment the sum of the values in registers I_7 and I_8 is equal to y. Any other path from state Q_3 into state Q_2 must include a transition from state Q_1 into state Q_2 and hence decreases the value in I_6 by 1. Thus, the number of sequences of the form (10.2.17) cannot be greater than x, and so the value in O_9 cannot be greater than xy.

When we have chess machines $\mathfrak{K}_1$ and $\mathfrak{K}_2$ computing functions $F_1(x_1, \dots, x_{m_1})$ and $F_2(y_1, \dots, y_{m_2})$, we can construct a machine $\mathfrak{K}$ computing the function

$$F(z_1, \dots, z_{m_1+m_2}) = F_1(z_1, \dots, z_{m_1}) + F_2(z_{m_1+1}, \dots, z_{m_1+m_2}). \tag{10.2.18}$$

For this purpose, we first renumber the states and registers of the machines $\mathfrak{K}_1$ and $\mathfrak{K}_2$, obtaining machines $\mathfrak{K}'_1$ and $\mathfrak{K}'_2$, respectively, in such a manner that:

(a) the number of the final state of the machine $\mathfrak{K}'_1$ is equal to the number of the initial state of machine $\mathfrak{K}'_2$, while all other numbers of states and registers of machines $\mathfrak{K}'_1$ and $\mathfrak{K}'_2$ are distinct;
(b) the final state of the machine $\mathfrak{K}'_2$ is the state Q_1;
(c) the output register of the machine $\mathfrak{K}'_1$ is the register O_4, and that of machine $\mathfrak{K}'_2$ is the register O_5;
(d) the numbers of all other states and registers (i.e., those not mentioned explicitly above) are greater than 6.

The instructions of the machine $\mathfrak{K}$ can be obtained by combining the instructions of the machines $\mathfrak{K}'_1$, $\mathfrak{K}'_2$, and the instructions (10.2.14). The initial state of machine $\mathfrak{K}$ coincides with that of machine $\mathfrak{K}'_1$, the input registers are those of machines $\mathfrak{K}'_1$ and $\mathfrak{K}'_2$, the output register is O_6, and the final state is Q_3.

Similarly, using the multiplication machine (10.2.15) instead of the addition machine (10.2.14), we can calculate $F_1(z_1, \dots, z_{m_1}) \cdot F_2(z_{m_1+1}, \dots, z_{m_1+m_2})$. So, we can calculate any polynomial that has a representation in which each variable occurs only once and every non-zero coefficient is equal to 1.

Let

$$D(x_1, \ldots, x_m) = 0 \tag{10.2.19}$$

be an arbitrary Diophantine equation. Our next goal is to construct chess machines $\mathfrak{K}_1'$ and $\mathfrak{K}_2''$ and initial positions of the knights such that inclusion (10.2.7) holds if and only if equation (10.2.19) has no solution.

Let A and B be polynomials with natural number coefficients such that

$$D^2(x_1, \ldots, x_m) = A(x_1, \ldots, x_m) - B(x_1, \ldots, x_m). \tag{10.2.20}$$

Thus, for any values of $x_1, \ldots, x_m$,

$$A(x_1, \ldots, x_m) \geq B(x_1, \ldots, x_m), \tag{10.2.21}$$

and if equation (10.2.19) has no solution, then

$$A(x_1, \ldots, x_m) \geq B(x_1, \ldots, x_m) + 1. \tag{10.2.22}$$

Furthermore, let $A'(y_1, \ldots, y_{mw})$ and $B'(z_1, \ldots, z_{mw})$ be polynomials with coefficients 0 and 1 such that

$$A(x_1, \ldots, x_m) = A'(\underbrace{x_1, \ldots, x_1}_{w \text{ times}}, \ldots, \underbrace{x_m, \ldots, x_m}_{w \text{ times}}), \tag{10.2.23}$$

$$B(x_1, \ldots, x_m) + 1 = B'(\underbrace{x_1, \ldots, x_1}_{w \text{ times}}, \ldots, \underbrace{x_m, \ldots, x_m}_{w \text{ times}}). \tag{10.2.24}$$

Finally, let $\mathfrak{K}_A$ and $\mathfrak{K}_B$ be chess machines computing A' and B' respectively. Without loss of generality, we will assume that the output register of the machine $\mathfrak{K}_B$ is B_{m+1} and the numbers of all other registers are greater than $m+9$.

Let us combine the instructions of machine $\mathfrak{K}_B$ with instructions of the form

$$\begin{aligned}
&Q_{m+2} \Longrightarrow Q_{m+3} R_{i_{1,1}} \ldots R_{i_{1,w}} X_1 \\
&\quad\vdots \\
&Q_{m+2} \Longrightarrow Q_{m+3} R_{i_{m,1}} \ldots R_{i_{m,w}} X_m \\
&Q_{m+3} \Longrightarrow Q_{m+2} \\
&Q_{m+3} \Longrightarrow Q_b,
\end{aligned} \tag{10.2.25}$$

where $R_{i_{1,1}}, \ldots, R_{i_{m,w}}$ are the input registers of the machine $\mathfrak{K}_B$, and Q_b is its initial state. The resulting machine will be denoted by $\mathfrak{K}_B''$. The initial state of machine $\mathfrak{K}_B''$ is Q_{m+2}, while its final state coincides with that of the machine $\mathfrak{K}_B$. The machine $\mathfrak{K}''_B$ will not be used for computing any function, so we are not specifying any input or output registers for it.

Beginning in state Q_{m+2} with all of its registers empty, the machine $\mathfrak{K}''_B$ first generates some values $x_1, \ldots, x_m$ in registers $X_1, \ldots, X_m$ and in the input registers of the machine $\mathfrak{K}_B$ and then calculates in register B_{m+1} the value of the polynomial $B(x_1, \ldots, x_m) + 1$; i.e., this register can contain any number up to and including the value of this polynomial.

Similarly, it is possible to construct a machine $\mathfrak{K}'_A$ that will calculate the value of $A(x_1, \ldots, x_m)$ in register A_{m+1}. If equation (10.2.19) has a solution, then machine $\mathfrak{K}''_B$ can have in its register B_{m+1} a value that is greater by 1 than the maximum value that the machine $\mathfrak{K}'_A$ can place in register A_{m+1} (with the correspnding values in the registers $X_1, \ldots, X_m$ satisfying equation (10.2.19)). On the other hand, if equation (10.2.19) has no solution, then the inequality (10.2.20) holds, and any value that the machine $\mathfrak{K}''_B$ can put in the register B_{m+1} can also be put in the register A_{m+1} by machine $\mathfrak{K}'_A$. However, this is not sufficient to decide whether (10.2.7) is true or false, because in (10.2.7) one has to compare the values in all the registers.

To overcome this obstacle, we are going to modify machines $\mathfrak{K}'_A$ and $\mathfrak{K}''_B$. In doing so we will assume, without loss of generality, that the numbering of the states and registers of these machines is such that no number is used to number a state or register of both machines except as follows: the numbers $1, \ldots, m$ are used to number the input registers $X_1, \ldots, X_m$ of both machines, the number $m+1$ is used to number the respective output registers A_{m+1} and B_{m+1} of the two machines, and the initial states Q_b are numbered b in both machines. Let Q_t be a state used neither by $\mathfrak{K}''_A$ nor by $\mathfrak{K}''_B$, where $t > m + 8$, and let Q_s be the final state of the machine $\mathfrak{K}''_A$. We extend the instructions of machine $\mathfrak{K}''_A$ by including the following groups of instructions:

(a) instructions of the form

$$Q_s R_i \Longrightarrow Q_t \tag{10.2.26}$$

for each register R_i of machine $\mathfrak{K}''_A$ different from $X_1, \ldots, X_{m+1}$;
(b) instructions of the form

$$Q_s \Longrightarrow Q_t R_j \tag{10.2.27}$$

for each register R_j of machine $\mathfrak{K}''_B$ different from $X_1, \ldots, X_{m+1}$;

(c) instructions of the form

$$Q_s \Longrightarrow Q_k \tag{10.2.28}$$

for every state Q_k of machine $\mathfrak{K}''_B$;
(d) the instruction

$$Q_t \Longrightarrow Q_s. \tag{10.2.29}$$

The resulting machine will be denoted by $\mathfrak{K}'$.

We will not add any new instructions to the machine $\mathfrak{K}''_B$, but we will add registers (actually, dummy registers) with numbers equal to the numbers of the states and registers of machine $\mathfrak{K}''_A$. The resulting machine will be denoted by $\mathfrak{K}''$.

For the initial square we take

$$\langle r_1, \ldots, r_n \rangle \tag{10.2.30}$$

where $r_{m+2} = 1$ and the remaining numbers in (10.2.30) are zero. Let us compare the sets of reachability $\mathfrak{K}'(r_1, \ldots, r_n)$ and $\mathfrak{K}''(r_1, \ldots, r_n)$. It was noted above that if equation (10.2.19) has a solution, then beginning with the same values in registers $X_1, \ldots, X_m$, machine $\mathfrak{K}''_B$, and hence machine $\mathfrak{K}''$ as well, can put in the register R_{m+1} a value that machine $\mathfrak{K}'_A$ cannot, and hence machine $\mathfrak{K}'$ cannot either. Thus, if equation (10.2.19) has a solution, inclusion (10.2.17) does not hold.

Now suppose that equation (10.2.19) has no solution and knight $\mathfrak{K}''$, having started from square (10.2.30), reached some square

$$\langle r'_1, \ldots, r'_n \rangle. \tag{10.2.31}$$

Let us check that knight $\mathfrak{K}'$, having started from the same square, also can reach square (10.2.31).

First of all, using instructions of the form (10.2.25) we place the values $r'_1, \ldots, r'_m$ in registers $X_1, \ldots, X_m$. Furthermore, we know that

$$r'_{m+1} \leq B(r_1, \ldots, r_m) + 1, \tag{10.2.32}$$

and hence by (10.2.22),

$$r'_{m+1} \leq A(r_1, \ldots, r_m). \tag{10.2.33}$$

So, using the instructions of the machine $\mathfrak{K}'_A$, we can put r'_{m+1} in R_{m+1}. After that, using instructions of the form (10.2.26) and (10.2.29), we can make all the registers of machine $\mathfrak{K}'_A$ empty, except for $X_1, \ldots, X_m$, and R_{m+1}. Then, using instructions of the form (10.2.27) and (10.2.29), corresponding coordinates of the square (10.2.31) can be placed in the registers of the machine $\mathfrak{K}'_B$ and, finally, using instructions of the form (10.2.28) and (10.2.29), we can reach the same state in which machine $\mathfrak{K}''$ reached square (10.2.31).

Thus, inclusion (10.2.7) holds if and only if equation (10.2.19) has no solution; i.e., Hilbert's Tenth Problem has been reduced to the problem of comparison of "chess power" for inclusion, so the latter problem is also undecidable.

Now we turn to comparison for equality. Clearly,

$$\mathfrak{K}'(r_1,\ldots,r_n) \supseteq \mathfrak{K}''(r_1,\ldots,r_n) \iff \mathfrak{K}'(r_1,\ldots,r_n) = \mathfrak{K}'(r_1,\ldots,r_n) \cup \mathfrak{K}''(r_1,\ldots,r_n). \quad (10.2.34)$$

Hence, it would be sufficient to manage to define a knight $\mathfrak{K}'''$ such that

$$\mathfrak{K}'''(r_1,\ldots,r_n) = \mathfrak{K}'(r_1,\ldots,r_n) \cup \mathfrak{K}''(r_1,\ldots,r_n). \quad (10.2.35)$$

Actually, we will define a knight $\mathfrak{K}'''$ that almost, but not quite, has property (10.2.35).

To unite the sets of reachability, we, roughly speaking, unite the set of rules of the two knights. If we did so in a straightforward manner, the resulting knight would be too powerful as a result of its ability to interlace moves using rules from the two sets. To avoid a mixture of this kind, we will use the following technique.

According to our agreement, machines $\mathfrak{K}'$ and $\mathfrak{K}''$ do not have registers and states with numbers between $m+4$ and $m+8$. Therefore, during the moves of the knights $\mathfrak{K}'$ and $\mathfrak{K}''$ from square (10.2.30), the corresponding coordinates remain equal to zero. Now we replace each rule (10.2.10) of knight $\mathfrak{K}'$ by a pair of rules

$$T_{m+4}R_{i_1}\ldots R_{i_a} \Longrightarrow T_{m+5}R_{j_1}\ldots R_{j_b} \quad (10.2.36)$$

and

$$T_{m+5}R_{i_1}\ldots R_{i_a} \Longrightarrow T_{m+4}R_{j_1}\ldots R_{j_b}. \quad (10.2.37)$$

Similarly, if (10.2.10) is a rule of knight $\mathfrak{K}''$, it is to be replaced by the rules

$$T_{m+6}R_{i_1}\ldots R_{i_a} \Longrightarrow T_{m+7}R_{j_1}\ldots R_{j_b} \quad (10.2.38)$$

and

$$T_{m+7}R_{i_1}\ldots R_{i_a} \Longrightarrow T_{m+6}R_{j_1}\ldots R_{j_b}. \quad (10.2.39)$$

Now we combine all the rules of the forms (10.2.36)–(10.2.39) and add four more rules

$$Q_{m+8} \Longrightarrow T_{m+4}Q_{m+2} \tag{10.2.40}$$
$$Q_{m+8} \Longrightarrow T_{m+5}Q_{m+2} \tag{10.2.41}$$
$$Q_{m+8} \Longrightarrow T_{m+6}Q_{m+2} \tag{10.2.42}$$
$$Q_{m+8} \Longrightarrow T_{m+7}Q_{m+2}. \tag{10.2.43}$$

The resulting knight will be denoted by $\mathfrak{K}_{AB}$. Consider its movements from a square

$$\langle p_1, \ldots, p_n \rangle \tag{10.2.44}$$

where $p_{m+8} = 1$ and all other coordinates are equal to 0. In its first move, the knight $\mathfrak{K}_{AB}$ chooses which of the knights to simulate. Namely, if the first move is by rule (10.2.40) or (10.2.41), then afterwards the registers T_{m+6} and T_{m+7} will always be empty, and of the two registers T_{m+4} and T_{m+5}, one will be empty and the other will contain 1, so subsequent moves of the knight $\mathfrak{K}_{AB}$ are to be according to rules (10.2.36)–(10.2.37). On the other hand, if the first move is according to rule (10.2.42) or (10.2.43), then all subsequent moves will be by rules (10.2.38)–(10.2.39).

Now we are in position to define knights $\mathfrak{K}'''$ and $\mathfrak{K}''''$ whose "chess power" is to be compared for exact equality.

The knight $\mathfrak{K}''''$ is permitted to make all the moves that are legal for the knight $\mathfrak{K}_{AB}$ plus two additional moves according to the rules

$$\begin{aligned} T_{m+4} &\Longrightarrow \\ T_{m+5} &\Longrightarrow \end{aligned} \tag{10.2.45}$$

These rules decrease the values in registers T_{m+4} and T_{m+5} without increasing any other values. Knight $\mathfrak{K}'''$ is allowed all the moves that are legal for the knight $\mathfrak{K}''''$ plus two additional moves according to the rules

$$\begin{aligned} T_{m+6} &\Longrightarrow \\ T_{m+7} &\Longrightarrow \end{aligned} \tag{10.2.46}$$

Let us compare the sets of squares that could be visited by knights $\mathfrak{K}'''$ and $\mathfrak{K}''''$ after starting from square (10.2.44). It is evident that knight $\mathfrak{K}'''$ can visit any square that can be visited by knight $\mathfrak{K}''''$. Now suppose that there is a square

$$\langle p'_1, \ldots, p'_n \rangle \tag{10.2.47}$$

that can be visited by knight $\mathfrak{K}'''$ but not by knight $\mathfrak{K}''''$. It is not difficult to see that this is possible only in the case where the first move was according to one of the rules (10.2.42)–(10.2.43) and the last move was according to one of the rules (10.2.46) not permitted to knight $\mathfrak{K}''''$. This means that during the remainder of its moves, knight $\mathfrak{K}'''$ simulated knight $\mathfrak{K}''$, and hence that the latter can also reach square (10.2.47) from square (10.2.30). On the other hand, knight $\mathfrak{K}''''$ cannot reach square (10.2.47), and hence this square cannot be visited from square (10.2.30) by the knight $\mathfrak{K}'$ either. Thus, inclusion (10.2.7) holds if and only if equality (10.2.8) holds, and hence the problem of comparison for exact equality is also undecidable.

Exercises

1. The Diophantine game with a non-computable winning strategy given in Section 10.1 was based on the example of a simple set constructed in Section 6.5. This example, in turn, was based on the existence of a universal Diophantine equation and the possibility of eliminating bounded universal quantifiers. (Universal equations were constructed in Chapter 4 and also in Section 6.4; various techniques for eliminating bounded universal quantifiers were presented in Sections 6.1–6.3.) The resulting Diophantine game (1.7) is defined by an extremely complicated polynomial. The entire construction can be simplified somewhat by eliminating the explicit construction of a Diophantine representation of a simple set (and taking advantage only of its existence). Consider the following game based on a universal equation

$$U_1(a, k, x_1, \dots, x_{\mathbf{m}}) = 0$$

(but not defined by it in the manner of (1.1)):

on the first move, player A chooses a number k;
on the second move, player B chooses a number b;
on the third move, player A chooses a number a;
player A is declared the winner if and only if the equation

$$U_1(a, k, x_1, \dots, x_{\mathbf{m}}) = 0$$

has no solution while the equation

$$U_1(b, k, y_1, \dots, y_{\mathbf{m}}) = 0$$

has a solution.

(a) Transform this game into the canonical form (1.1).
(b) Show that B has a winning strategy.
(c) Show that B cannot implement this winning strategy effectively.

2. Consider the class of programs without loops and conditional branches consisting of instructions of the form $x \leftarrow 1,\ x \leftarrow x{+}y,\ x \leftarrow (x \operatorname{div} y)$. Show that the problem of determining, given such a program, whether it computes the function identically equal to 1 is undecidable.

Commentary

Diophantine games were introduced in Jones [1974]. There he used some ideas from Rabin [1957], where similar games were treated in the more general context of computability theory. There was no difficuly in carrying these ideas over to Diophantine games once it had been shown that semidecidable sets are Diophantine. The main achievement in Jones [1982a] consisted not in pure existence theorems like those in Section 10.1, but in the possibility of exhibiting specific polynomials defining Diophantine games for which particular decision problems are undecidable. Notably, in this paper the equation

$$\begin{aligned}
&\Big\{\{a_1 + a_6 + 1 - b_4\}^2 \cdot \Big\{\big\langle (a_6 + a_7)^2 + 3a_7 + a_6 - 2b_4\big\rangle^2 \\
&\quad + \Big\langle \big[(b_9 - a_7)^2 + (b_{10} - a_9)^2\big]\big[(b_9 - a_6)^2 + (b_{10} - a_8)^2((b_4 - a_1)^2 \\
&\quad + (b_{10} - a_9 - b_1)^2)\big]\big[(b_9 - 3b_4)^2 + (b_{10} - a_8 - a_9)^2\big]\big[(b_9 - 3b_4 - 1)^2 \\
&\quad + (b_{10} - a_8a_9)^2\big] - a_{12} - 1\Big\rangle^2 + \big\langle [b_{10} + a_{12} + a_{12}b_9a_4 - a_3]^2 \\
&\quad + [b_5 + a_{13} - b_9a_4]^2\big\rangle\Big\} - b_{13} - 1\Big\}\{a_1 + b_5 + 1 - a_5\}\Big\{\big\langle (b_5 - b_6)^2 \\
&\quad + 3b_6 + b_5 - 2a_5\big\rangle^2 + \Big\langle \big[(a_{10} - b_6)^2 + (a_{11} - b_8)^2\big]\big[(a_{10} - b_5)^2 \\
&\quad + (a_{11} - b_7)^2((a_5 - a_1)^2 + (a_{11} - b_8 - a_2)^2)\big]\big[(a_{10} - 3a_5)^2 \\
&\quad + (a_{11} - b_7 - b_8)^2\big]\big[(a_{10} - 3a_5 - 1)^2 + (a_{11} - b_7b_8)^2\big] - b_{11} - 1\Big\rangle^2 \\
&\quad + \big\langle [a_{11} + b_{11} + b_{11}a_{10}b_3 - b_2]^2 + [a_{11} + b_{12} - a_{10}b_3]^2\big\rangle\Big\} = 0
\end{aligned}$$

was presented as an example of a Diophantine game in which the second player has a winning strategy but no computable winning strategy. The proof was based on an idea somewhat more complicated than that of Exercise 10.1. For other complexity questions connected with Diophantine games, see Jones [1991].

In Section 10.2 we used chess terminology to make the presentation more visually intuitive. However, the problems treated there are known in the literature under a different terminology. In the first place, we should mention *systems of vector addition*, which are adequate for our generalized knights. Due to the fact that our knights satisfy the inequalities $|\delta_{i,j}| \leq 1$ (recall rules (10.2.4)), the knights' moves can easily be simulated by *Petri nets*, and our results can readily be carried over to this formalism. Systems of vector addition and Petri nets were originally introduced as tools for investigations of parallel computation.

The undecidability of comparison for inclusion was established by Michael Rabin in 1966 (see Hack [1976]). At that time the undecidability of Hilbert's Tenth Problem had not yet been established, so Rabin used a weaker result, the undecidability of exponential Diophantine Equations (see the Commentary to Chapter 5 for the history of this result), and therefore had to show not only that addition and multiplication can be calculated by a chess machine, but also that the same was true of exponentiation. In 1972, Rabin gave a simplified proof based on the undecidability of Hilbert's Tenth Problem (see Hack [1976]). Rabin published neither proof. In Section 10.2 we followed Hack [1976], where the stronger result, the undecidability of comparison for equality, was also established.

Hilbert's Tenth Problem has been used to establish the undecidability of many other decision problems. It is impossible to survey all the applications in this book, particularly because some of them present technical difficulties or require special knowledge in the relevant areas. Here we confine ourselves to giving references to the literature.

Hilbert's Tenth Problem was used for investigations of various decision problems arising in algebra by Aĭvazyan [1983, 1988], Anick [1985], Bokut' [1972], Durnev [1974], Huber-Dyson [1984], Kharlampovich [1989a, 1989b], Kleĭman [1979], Lyndon and Schupp [1977], Miller [1973], Myasnikov and Remeslennikov [1986], Remeslennikov [1979], Repin [1983], Roman'kov [1977, 1979], Rozenblat [1985], Sapir [1989], Sapir and Kharlampovich [1989], Valiev [1975], and Vazhenin [1987, 1988, 1989].

The equivalence of "semidecidable" and "Diophantine" has found application in model theory and proof theory; see Adler [1969b], Adamowicz [1992], Baur [1974], Baxter [1978], Calude and Păun [1983], Cantone, Cutello, and Policrito [1990],

Carstens [1977], Chaitin [1974, 1987], Davis [1989], Dyson, Jones, and Shepherdson [1982], Friedrichsdorf [1976], Gaifman [1975], Gaifman and Dimitracopoulos [1982], Germano [1976], Goldfarb [1981], Guaspari [1979], Hatcher and Hodgson [1981], Hirschfeld [1975], Hirschfeld and Wheeler [1975], Joseph and Young [1981], Kaye [1987, 1990], Kaye, Paris, and Dimitrocopoulos [1988], Kosovskiĭ [1967], Lipshitz [1979], Livesey, Siekmann, Szabó, and Unvericht [1979], Macintyre and Simmons [1973], Manevitz [1976a, 1976b], Máté [1990], McAloon [1978], Mints [1975], Mlček [1976], Motohashi [1984], Pacholski and Szwast [1991], Paris and Dimitrocopoulos [1982], Penzin [1979], A. Robinson [1973], J. Robinson [1973], Siekmann and Szabó [1989], Singer [1978], Tiden and Arnborg [1987], Wilkie [1975, 1980], and Wilmers [1985].

The undecidability of Hilbert's Tenth Problem has also been useful in theoretical computer science; see Beck [1975], Broy, Wirsing, and Pepper [1987], Chan [1987], Clarke and Richardson [1985], Cohen [1990], Dańko [1980, 1981], Gorskiĭ and Yakubovich [1985], Gurari [1985], Gurari and Ibarra [1982], Hack [1976], Harel, Pnueli, and Stavi [1983], Heering [1986], Howell, Rosier, Huynh, and Yen [1986], Huet and Oppen [1980], Ibarra [1982], Ibarra and Leininger [1980, 1982, 1983], Kasami and Tokura [1971], Kucherov [1988], Laski [1986], Mayr and Meyer [1981], Meyer auf der Heide [1989], Reif and Lewis [1986], Rosier [1986], Shiryaev [1989], Statman [1981], Stotskiĭ [1973, 1971a, 1971b], and Turakainen [1978, 1981].

The undecidability of Hilbert's Tenth Problem implies the impossibility of the existence of algorithms for certain problems of mathematical programming; see Durnev and Korablëva [1988], Jeroslow [1973], Maňas [1971], and Tarasov and Khachiyan [1980].

We conclude by mentioning a paper by Goodstein [1975] that is quite different from most research using the undecidability of Hilbert's Tenth Problem. In this paper, a result that does not at all concern the existence of algorithms is obtained as a corollary of two results that do, the undecidability of Diophantine equations and the decidability of equations of a special form, established in Goodstein and Lee [1966].

Appendix

1 The Four Squares Theorem

THEOREM (Lagrange [1772]). The Diophantine equation

$$x_1^2 + x_2^2 + x_3^2 + x_4^2 = a \tag{1.1}$$

is solvable in x_1, x_2, x_3, x_4 for any non-negative a.

It follows from the *Euler Identity*

$$\begin{aligned}(x_1^2 + x_2^2 + x_3^2 + x_4^2)&(y_1^2 + y_2^2 + y_3^2 + y_4^2)\\ &= (x_1y_1 + x_2y_2 + x_3y_3 + x_4y_4)^2 + (x_1y_2 - x_2y_1 + x_3y_4 - x_4y_3)^2\\ &\quad + (x_1y_3 - x_3y_1 + x_2y_4 - x_4y_2)^2 + (x_1y_4 - x_4y_1 + x_2y_3 - x_3y_2)^2 \end{aligned} \tag{1.2}$$

that it is sufficient to prove the theorem for the special case when a is an odd prime.

Let us fix such a value of a and consider the equation

$$x_1^2 + x_2^2 + x_3^2 + x_4^2 = ab \tag{1.3}$$

with a single parameter b and four unknowns x_1, x_2, x_3, x_4. First, we show that equation (1.3) has a solution for some $b = b_0$, where

$$0 < b_0 < a. \tag{1.4}$$

Second, we show that if equation (1.3) has a solution for $b = b'$, where

$$1 < b' < a, \tag{1.5}$$

then there is another number b'' such that

$$0 < b'' < b' \tag{1.6}$$

and equation (1.3) has a solution for $b = b''$. Clearly, this will prove the theorem.

It is easy to check that all $(a+1)/2$ numbers

$$0^2, 1^2, \ldots, \left(\frac{a-1}{2}\right)^2 \tag{1.7}$$

are pairwise incongruent modulo a. Similarly, so are the $(a+1)/2$ numbers

$$-1-0^2, -1-1^2, \ldots, -1-\left(\frac{a-1}{2}\right)^2. \tag{1.8}$$

The lists (1.7) and (1.8) together contain $a+1$ numbers; hence, they contain two numbers congruent modulo a; i.e., there are numbers x_1, x_2, and b_0 such that

$$0 \le x_1 \le \frac{a-1}{2}, \qquad 0 \le x_2 \le \frac{a-1}{2} \tag{1.9}$$

and

$$x_1^2 + x_2^2 + 1 = ab_0. \tag{1.10}$$

The first inequality in (1.4) is evident, and the second can now easily be deduced from (1.9).

Now let

$$x_1^2 + x_2^2 + x_3^2 + x_4^2 = ab', \tag{1.11}$$

where b' satisfies (1.5). Furthermore, let y_1, y_2, y_3, y_4 be numbers such that

$$y_i \equiv x_i \pmod{b'}, \tag{1.12}$$

$$-b'/2 < y_i \le b'/2. \tag{1.13}$$

By (11) and (12),

$$y_1^2 + y_2^2 + y_3^2 + y_4^2 = b'b'' \tag{1.14}$$

for some b''. It follows from (13) that $b'' \le b'$. Equality is possible here only if $y_1 = y_2 = y_3 = y_4 = b'/2$. But then $|x_1| = |x_2| = |x_3| = |x_4| = b'/2$ and $b' = a$, which contradicts (1.5). Thus, the second inequality in (1.6) is valid. If $b' = 0$, then $y_1 = y_2 = y_3 = y_4 = 0$ and $b' \mid a$, in contradiction with (1.5) and the assumption that a is a prime. So, the first inequality in (1.6) holds as well.

Multiplying (11) and (14) and using the Euler Identity, we see that ab'^2b'' can be represented as the sum of four squares from the right-hand side in (1.2). It follows from (12) and (11) that each of these four numbers is divisible by b'^2; after division we obtain a solution of equation (1.3), as desired, for $b = b''$.

2 Chinese Remainder Theorem

THEOREM. Let $q_1, \ldots, q_n$ be pairwise coprime numbers, and let $a_1, \ldots, a_n$ be numbers such that

$$0 \le a_i < q_i, \qquad i = 1, \ldots, n. \tag{2.1}$$

Then there is exactly one number a such that

$$a_i = \operatorname{rem}(a, q_i), \qquad i = 1, \ldots, n \tag{2.2}$$

and

$$0 \le a < q_1 \cdots q_n. \tag{2.3}$$

To prove this, consider the $q_1 \cdots q_n$ n-tuples

$$\begin{array}{c} \langle \operatorname{rem}(0, q_1), \ldots, \operatorname{rem}(0, q_n) \rangle, \\ \vdots \\ \langle \operatorname{rem}(q_1 \cdots q_n - 1, q_1), \ldots, \operatorname{rem}(q_1 \cdots q_n - 1, q_n) \rangle. \end{array} \tag{2.4}$$

They all are different because if

$$\operatorname{rem}(a', q_i) = \operatorname{rem}(a'', q_i), \tag{2.5}$$

then

$$q_i \mid a' - a'' \tag{2.6}$$

and, by the assumption that $q_1, \ldots, q_n$ are pairwise prime,

$$q_1 \cdots q_n \mid a' - a'', \tag{2.7}$$

which is possible only if $a' = a''$.

Clearly, there are exactly $q_1 \cdots q_n$ n-tuples $\langle a_1, \ldots, a_n \rangle$ satisfying (2.1), and every n-tuple $\langle a_1, \ldots, a_n \rangle$ among the $q_1 \cdots q_n$ n-tuples (2.4) satisfies (2.1). Therefore, each n-tuple satisfying (2.1) is listed in (2.4) exactly once.

3 Kummer's Theorem

THEOREM (Kummer [1852])[1]. If p is a prime number, then its exponent in the canonical expansion of the binomial coefficient $\binom{m+n}{m}$ into prime factors is equal to the the number of carries required when adding the numbers m and n represented in the base p.

To prove this, note that the equation

$$\binom{m+n}{m} = \frac{(m+n)!}{m!\,n!} \tag{3.1}$$

[1]See also Singmaster [1974].

implies that

$$\deg_p\left(\binom{m+n}{m}\right) = \deg_p((m+n)!) - \deg_p(m!) - \deg_p(n!), \tag{3.2}$$

where $\deg_p(k)$ is the exponent of p in the prime factorization of k. It is not difficult to see that

$$\deg_p(k!) = k \operatorname{div} p + k \operatorname{div} p^2 + \cdots, \tag{3.3}$$

because among the numbers 1, ..., k, there are exactly $k \operatorname{div} p$ numbers divisible by p, exactly $k \operatorname{div} p^2$ numbers divisible by p^2, and so on. Thus,

$$\deg_p\left(\binom{m+n}{m}\right) = \sum_{k\geq 1}\left((m+n) \operatorname{div} p^k - m \operatorname{div} p^k - n \operatorname{div} p^k\right). \tag{3.4}$$

Now it suffices to note that in this sum, the kth summand is equal to either 1 or 0, depending on whether or not there is a carry from the $(k-1)$th digit.

4 Summation of a generalized geometric progression

Let

$$G_n(q,w) = \sum_{i=0}^{w} q^i i^n \tag{4.1}$$

be the sum of a generalized geometric progression. For an ordinary geometric progression, i.e., when $n = 0$, we have

$$G_0(q,w) = \frac{q^{w+1}-1}{q-1} \tag{4.2}$$

for $q \neq 1$. We show by induction that for all n, there are polynomials P_n with integer coefficients such that

$$G_n(q,w) = \frac{P_n(q, q^w, w)}{(q-1)^{n+1}}. \tag{4.3}$$

Replacing w by $w+1$ in (4.1) and differentiating with respect to q, we obtain

$$\begin{aligned}
\frac{\partial}{\partial q} G_n(q, w+1) &= \sum_{i=0}^{w+1} q^{i-1} i^{n+1} \\
&= \sum_{i=0}^{w} q^i (i+1)^{n+1} \\
&= \sum_{i=0}^{w} \sum_{j=0}^{n+1} \binom{n+1}{j} q^i i^j \\
&= \sum_{j=0}^{n+1} \binom{n+1}{j} G_j(q, w).
\end{aligned} \tag{4.4}$$

Now the polynomials P_n can be obtained using the recurrent relation

$$G_{n+1}(q, w) = \frac{\partial}{\partial q} G_n(q, w+1) - \sum_{j=0}^{n} \binom{n+1}{j} G_j(q, w). \tag{4.5}$$

Hints to the Exercises

1 Principal Definitions

1. Clearly, equation (1.3.2) is solvable in natural numbers if and only if the equation $D(x_1-1,\ldots,x_m-1)=0$ is solvable in positive integers. Similarly, equation (1.3.2) is solvable in positive integers if and only if the equation $D(x_1+1,\ldots,x_m+1)=0$ is solvable in natural numbers.

2. It is known (see, for example, Dickson [1939]) that every natural number of the form $4x+1$ is the sum of the squares of three integers. Clearly, two of these are even and the third is odd, so x has the representation $x=y_1^2+y_2^2+y_3^2+y_3$, which enables one to achieve a merely threefold increase in the number of unknowns. It is also known (see, for example, Andrews [1986]) that every natural number can be represented as the sum of three triangular numbers: $y_1(y_1-1)/2+y_2(y_2-1)/2+y_3(y_3-1)/2$. This provides another method for achieving a threefold increase.

The so-called Pell equation $y^2-dz^2=1$ is solvable in positive y and z for every d that is not a perfect square. This fact allows one to replace the last m equations in system (1.3.3) by $y_i^2-(4x_i-1)z_i^2=1$ and add the equation

$$2z_1\cdots z_m=w_1(w_1-1)+w_2(w_2-1)+w_3(w_3-1)+2.$$

This limits the increase in the number of unknowns to only $3m+3$. However, using Exercise 8.2, one can eliminate $y_1, \ldots, y_m$ at the cost of introducing yet another unknown, thus obtaining an equation in $2m+4$ unknowns. This bound was improved to $2m+2$ by Zhi-Wei Sun [1992].

The powerful technique presented in Chapter 4 will enable us to prove the existence of a constant **m** such that any equation can be transformed into an equivalent equation with **m** integer unknowns (see also Section 8.1).

3. Instead of (1.3.6), one can use the equation

$$\prod D(\pm x_1,\ldots,\pm x_m)=0,$$

where the product is taken over all 2^m possible choices of the plus and minus signs.

4. Instead of equation (1.4.8), one may use the equation

$$\begin{aligned}&(\chi_{0,1}^2+\chi_{0,2}^2+\chi_{0,3}^2+\chi_{0,4}^2+1)\\&\cdot(1-D^2(\chi_{0,1}^2+\chi_{0,2}^2+\chi_{0,3}^2+\chi_{0,4}^2,\ldots,\chi_{m,1}^2+\chi_{m,2}^2+\chi_{m,3}^2+\chi_{m,4}^2))-1=a.\end{aligned}$$

5. Suppose that the polynomial $P(x_1, \dots, x_m)$ assumes only prime values. Let $p = |P(1, \dots, 1)|$ and consider the polynomial $Q(y_1, \dots, y_m) = P(py_1 + 1, \dots, py_m + 1)$. Clearly,

$$Q(y_1, \dots, y_m) \equiv P(1, \dots, 1) \equiv 0 \pmod p,$$

and hence $Q(y_1, \dots, y_m) = \pm p$. Thus Q is a constant, and so is P as well.

6. Considering the case in which the unknowns range over the natural numbers, we can use the following equation instead of equation (1.4.8):

$$\begin{aligned} x_0\bigl(1 - x_0 D^2(x_0 - 1, x_1, \dots, x_m)\bigr) - 1 \\ - (x_0 + 1)x_{m+1}\bigl(1 + x_0 D^2(x_0 - 1, x_1, \dots, x_m)\bigr) = a. \end{aligned}$$

7. The set defined by equation (1.4.7) is exactly the set of all integer values of the fraction

$$\frac{x_0 + D^2(x_0, \dots, x_m)}{1 + (x_0 + 1)D^2(x_0, \dots, x_m)}.$$

8. (a) For A one can take any polynomial that assumes each of its values only once (for examples of such polynomials, see Section 3.1 and Exercise 3.1). For a counterpart of equation (1.4.8), one can take the equation

$$\bigl(A(y_1, \dots, y_n) + 1\bigr)\bigl(1 - D^2(y_1, \dots, y_n, x_1, \dots, x_m)\bigr) - 1 = A(a_1, \dots, a_n),$$

where $D(a_1, \dots, a_n, x_1, \dots, x_m) = 0$ is an equation defining the set $\mathfrak{M}$.
(b) For such a system one can take

$$\begin{aligned} (y_1 + 1)\bigl(1 - D^2(y_1, \dots, y_n, x_1, \dots, x_m)\bigr) - 1 &= a_1 \\ y_2 &= a_2 \\ &\vdots \\ y_m &= a_m. \end{aligned}$$

9. One can choose

$$E(b_1, \dots, b_n, x_0, \dots, x_m) = (x_0 + 1)\bigl(1 - D^2(x_0, b_1, \dots, b_n, x_1, \dots, x_m)\bigr) - 1,$$

where D is the polynomial from representation (1.5.5).

10. Here are the corresponding representations:

(a) $(x+2)(y+2)=a$;
(b) $x(2y+3)=a$;
(c) $x^2 < a < (x+1)^2$.

11. This was originally proved by J. Robinson [1952]. Another representation can easily be obtained using the technique of Chapter 2.

12. One of the necessary and sufficient conditions for p/q and r/s to be consecutive convergents consists in their satisfying the equation $|ps - qr| = 1$ together with one of the two two-way inequalities $p/q \leq \chi \leq r/s$ or $r/s \leq \chi \leq p/q$. A number χ can be specified by rational numbers a/b and c/d and a polynomial $A(x)$ with integer coefficients such that χ is the unique zero of the equation $A(\eta) = 0$ in the interval $[a/b, c/d]$ and $A(a/b) < 0$, $A(c/d) > 0$. Then

$$x/y < \chi \iff x/y < a/b \vee [x/y < c/d \,\&\, A(x/y) < 0].$$

13. Trigonometric Diophantine equations were introduced by Conway and Jones [1976] in what, formally speaking, is a more general setting than was given in Exercise 1.11 and were there reduced to Hilbert's Tenth Problem.

14. See, for example, Amice [1964].

15. This result of Alfred Tarski was published in J. Robinson [1952]. The proof was based on a generalization, discovered by Tarski [1951], of Sturm's theorem about the number of real zeros of a polynomial in a given interval. Another proof can easily be obtained by using the technique of Chapter 6 for eliminating bounded universal quantifiers, together with trivial bounds for the zeros in terms of the coefficients.

2 Exponentiation Is Diophantine

1. This was proved by J. Robinson [1952].

2. This was also proved by J. Robinson [1952]. If the 4-tuple $\langle a_1, b_1, a_2, b_2 \rangle$ belongs to the set of Exercise 2.1, $b_2 = b_1 + 1$, and b_1 is large enough, then

$$n = ((a_2 - a_1)b_1) \operatorname{div} a_1.$$

3. This was proved by Matiyasevich [1968b] by expanding quadratic irrationalities of the form

$$\sqrt{\left((2b+1)^m+b+1\right)^2-2b-1}$$

into continued fractions.

4. The values of x and z are related by $z = x - ay/2$.

5. These sequences were used, in particular, in Matiyasevich [1972a]. For these sequences, the equation

$$x^2 - bxy - y^2 = \pm 1$$

serves as a characteristic equation analogous to equation (2.1.12).

6. For example, one could use the congruence (2.1.5), since for sufficiently large d,

$$\alpha_b(n) = \mathrm{rem}(\alpha_{b+d}(n), d).$$

7. A proof can easily be carried out using induction on n. For sufficiently large x,

$$b^c = \mathrm{rem}\big((x-b)\alpha_x(c) - \alpha_x(c+1), bx - b^2 - 1\big).$$

The idea for this Diophantine representation originated in J. Robinson [1952], where a similar congruence was given for the solutions of the Pell equation from Exercise 1.2.

8. This was done by J. Robinson [1952].

9. It was shown by J. Robinson [1952] that if we know the lowest possible height of the tower of exponents, then we can find a generalized Diophantine representation for some relation of exponential growth, after which it suffices to use Exercise 2.8.

10. This was proved by Davis [1968] under the more restricted assumption that the solution $u = r = 1, \;\; v = s = 0$ is unique. Weakening this assumption to the existence of only finitely many solutions presents no difficulty.

3 Diophantine Coding

1. This polynomial was suggested for this purpose by Kosovskiĭ [1971].

2. According to Bollman and Laplaza [1978], this was established by Lew and Rosenberg [1975]. See also Lew and Rosenberg [1978a, 1978b].

3. The fact that this function is Diophantine was first established by Davis, Putnam, and Robinson [1961] on the base of the following identity:

$$\prod_{k\leq c}(a+bk) = \binom{a/b+c}{c+1} b^{c+1}(c+1)!$$

Another proof, which does not require first showing that binomial coefficients with rational arguments are Diophantine, was given by J. Robinson [1971]. The proof was based on the congruence

$$\prod_{k\leq c}(a+bk) \equiv \binom{x+c}{c+1} b^{c+1}(c+1)! \pmod{m},$$

which is valid when $a \equiv bx \pmod{m}$.

4. See Jones and Matiyasevich [1982b]. For sufficiently large x, the integer part of $x^n/\sqrt{1-4/x}$ is the cipher to the base x of a tuple whose first element is $\binom{2n}{n}$.

5. If $2 < m$, $q_m < m^m$, and $m^{2m} < t$, then the inequality

$$\left| t^m e^{1/t} - \frac{p_m}{q_m} \right| < \frac{1}{t}$$

is equivalent to the equation

$$\frac{p_m}{q_m} = \sum_{k=0}^{m} \frac{t^{m-k}}{k!}.$$

The expression $e^{1/t}$ may be replaced by $\left(1+\frac{1}{yt}\right)^y$ for sufficiently large y, so it suffices to note that

$$m! = \left(\frac{p_m}{q_m} - \frac{tp_{m-1}}{q_{m-1}}\right)^{-1}.$$

6. Use the identity

$$\sum_{k=0}^{\infty} \frac{B_k}{k!}\tau^k = \frac{\tau}{e^\tau - 1},$$

similar to the one in Exercise 3.5.

7. Take advantage of the fact that according to the Staudt-Klausen theorem,

$$B_{2m} = Z - \sum \frac{1}{p},$$

where Z is an integer and the summation is over all primes p such that $(p-1) \mid 2m$.

8.
$$\begin{aligned}
\neg\,\mathrm{Code}(a,b,c) &\iff b < 2 \vee a \geq b^c \\
\neg\,\mathrm{Equal}(a_1,b_1,c_1,a_2,b_2,c_2) &\iff \neg\,\mathrm{Code}(a_1,b_1,c_1) \vee \neg\,\mathrm{Code}(a_2,b_2,c_2) \vee c_1 \neq c_2 \\
&\qquad \vee \exists k\,[\mathrm{Elem}(a_1,b_1,k) \neq \mathrm{Elem}(a_2,b_2,k)] \\
\neg\,\mathrm{NotGreater}(a_1,b_1,a_2,b_2) &\iff \exists k\,[\mathrm{Elem}(a_1,b_1,k) > \mathrm{Elem}(a_2,b_2,k)] \\
\neg\,\mathrm{Small}(a,b,c,e) &\iff \neg\,\mathrm{Code}(a,b,c) \vee \exists k\,[\mathrm{Elem}(a,b,k) > e]
\end{aligned}$$

9.
$$\begin{aligned}
a = \mathrm{Mult}(p,q,r,u,v,w) \iff \exists xy\,[a = xy \;\&\; &\mathrm{Equal}(p,q,r,u,(qv)^w,r) \\
\&\; &\mathrm{Equal}(u,v,w,y,qv,w)]
\end{aligned}$$

$$\begin{aligned}
&\mathrm{Add}(p,q,r,u,v,w) \\
&\quad = \mathrm{Mult}(p,q,r,\mathrm{Repeat}(1,v,w),v,w) + \mathrm{Mult}(\mathrm{Repeat}(1,q,r),q,r,u,v,w)
\end{aligned}$$

10. $s = \mathrm{Sum}(a,b,c) \iff \exists x\,[\mathrm{Equal}(a,b,c,x,bc+2,c) \;\&\; s = \mathrm{rem}(x,bc+1)]$

11.
$$\begin{aligned}
l = \mathrm{Less}(a_1,a_2,b) \iff \exists w x_1 x_2 y\,[\,&\mathrm{Small}(l,b,w,1) \\
\&\; &\mathrm{Equal}(a_1,b,w,x_1,2b,w) \\
\&\; &\mathrm{Equal}(a_2,b,w,x_2,2b,w) \\
\&\; &\mathrm{Equal}(l,b,w,y,2b,w) \\
\&\; &\mathrm{NotGreater}(x_1+y,2b,x_2,2b) \\
\&\; &\mathrm{NotGreater}(x_2,2b,x_1+by,2b)]
\end{aligned}$$

12. Without loss of generality, we will assume that $a \geq 2$. Let $b = a+1$, $c = a-1$, $y = \mathrm{Repeat}(1,b,c)$, and let z be the cipher of the tuple $\langle 2,3,\dots,a\rangle$ of length c to the base b. (This cipher can be explicitly expressed in terms of a, b, and c; see the Appendix.) The tuple with cipher $x = \mathrm{Mult}(z,b,c,z,b,c)$ contains all composite numbers that are not greater than a. A number q, also not exceeding a, is prime if and only if

$$\mathrm{NotEqual}\big(x,\mathrm{Repeat}(q,b^2,c^2),c^2\big) = w,$$

where

$$\text{NotEqual}(a_1, a_2, d) = \text{Less}(a_1, a_2, d) + \text{Less}(a_2, a_1, d),$$
$$w = \text{Repeat}(1, b^2, c^2).$$

Put $u = \text{NotEqual}\big(\text{Mult}(y, b, c, x, b^2, c^2), \text{Mult}(z, b, c, w, b^2, c)\big)$. In the tuple with code $\langle u, b^{3c}, c\rangle$, exactly $\pi(a)$ elements are equal to w, so

$$\pi(a) = c - \text{Sum}\Big(\text{NotEqual}\big(u, \text{Repeat}(w, b^{3c^2}, c), b^{3c^2}\big), b^{3c^2}, c)\Big).$$

13. A number p is the nth prime if and only if $\pi(p-1) = n-1,\ \ \pi(p) = n$. It suffices to use Exercises 3.7 and 1.9.

14. Take advantage of the fact that the binary representation of a number can be obtained from its 2^d-ary notation by replacing each of the digits by its binary representation (padded, when required, by leading zeros). This allows one to apply Kummer's Theorem to obtain a Diophantine representation for the 4-place relation $\text{Small}(a, 2^d, c, 2^f)$.

15. This was proved by Matiyasevich [1971b]. With respect to a sufficiently large base w, the cipher d of the tuple $\langle \delta(0), \dots, \delta(n)\rangle$ satisfies the congruence

$$(b_0 w^k + \dots + b_{k-1} w) d \equiv a_0 + \dots + a_{k-1} w^{k-1} + d \pmod{w^{n+1}}$$

and the inequality $d < w^{n+1}$ and is uniquely determined by these two conditions.

16.

$$\begin{aligned} m = \text{Min}(a_1, a_2, b) \iff & \text{NotGreater}(m, b, a_1, b)\ \&\ \text{NotGreater}(m, b, a_2, b) \\ & \&\ \text{Less}\big(\text{Less}(m, a_1, b), \text{Less}(m, a_2, b)\big) = \text{Less}(m, a_2, b) \\ & \&\ \text{Less}\big(\text{Less}(m, a_2, b), \text{Less}(m, a_1, b)\big) = \text{Less}(m, a_1, b) \end{aligned}$$

17. The existence and the uniqueness of this representation was established by Nielsen [1924]. Concatenation corresponds to matrix multiplication.

4 Universal Diophantine Equations

1. Consider the equation

$$U^2(a_1, \dots, a_{n-1}, x_0, \dots, x_l, y_1, \dots, y_w) + (a_n - 2^{2^l}\, \text{Cantor}_{l+1}(x_0, \dots, x_l))^2 = 0.$$

Let $k_1^D, \dots, k_l^D$ be chosen so that this equation is solvable in $x_0, \dots, x_l, y_1, \dots, y_w$ precisely when equation (4.1.3) is. Then the equation

$$U(a_1, \dots, a_{n-1}, 2^{2^l} \operatorname{Cantor}(a_n, k), k_1^D, \dots, k_l^D, y_1, \dots, y_w) = 0$$

is also universal.

2. Universal equations of the following forms were given by Matiyasevich [1972b]:

$$V(a_1, \dots, a_n, x_1, \dots, x_m) = k,$$
$$V(k, x_1, \dots, x_m) = W(a_1, \dots, a_n),$$

and

$$V(x_1, \dots, x_m) = W(a_1, \dots, a_n, k).$$

3. $\operatorname{Format}(d, e, f) \iff d > 1 \,\&\, \operatorname{Equal}(f + 2, 2, d^e + 1, f + 2^{d^{e+1}}, 2^d, d^e + 1)$

4. The idea of a universal equation with "transparent" structure was proposed by Kosovskiĭ [1971]. Possible values of d and m were determined, in particular, by Jones [1982b] (see Commentary to Chapter 8).

5 Hilbert's Tenth Problem Is Unsolvable

1.

```
DOUBLE = LAST; MARK(3);
           while THEREWAS(3) do INC od
```

2.

```
ERASE = STAR; RIGHT; WRITE(2); MARK(2);
    while THEREIS(2) do
        while READ(2) do RIGHT od;
        while READ(0) do LEFT; WRITE(0); RIGHT; WRITE(2) od;
        while READ(1) do LEFT; WRITE(1); RIGHT; WRITE(2) od;
        while READ(λ) do
            while LEFT; READ(2) do WRITE(λ) od
        od
    od
```

3.
$$
\begin{aligned}
\text{BIN}(1) = \text{NEW};\ & \\
& \textbf{while } \text{READNOT}(,) \textbf{ do} \\
& \quad \text{STAR; RIGHT;} \\
& \quad \textbf{while } \text{READ}(2) \textbf{ do } \text{RIGHT } \textbf{od}; \\
& \quad \textbf{while } \text{READ}(0) \textbf{ do } \text{WRITE}(2); \text{DOUBLE } \textbf{od}; \\
& \quad \textbf{while } \text{READ}(1) \textbf{ do } \text{WRITE}(2); \text{DOUBLE; INC } \textbf{od} \\
& \textbf{od}; \\
& \text{WRITE}(2); \text{ERASE}
\end{aligned}
$$

$$\text{BIN}(n+1) = \text{BIN}(1); \text{BIN}(n)$$

4. This can be done in two ways. The first is to construct a Turing machine that transforms the unary representation of a tuple into the binary representation. The second is to modify condition (5.5.38), replacing it by

$$\langle t, b, a\rangle = \langle \kappa, b, 1\rangle + \langle d_1, b, c_1\rangle + \langle \rho, b, 1\rangle + \cdots + \langle d_n, b, c_n\rangle + \langle \rho, b, 1\rangle,$$

where ρ is the number of the symbol ",", and the numbers $d_1, \ldots, d_n, c_1, \ldots, c_n$ satisfy the conditions $\text{Equal}(a_1, 2, c_1, d_1, b, c_1), \ldots, \text{Equal}(a_n, 2, c_n, d_n, b, c_n)$.

5. In order to prove that a Turing-computable function is Diophantine, it is sufficient to extract its value from the tuple whose cipher is $\text{AfterT}(k, p, t)$, where k, p, and t are defined by conditions (5.5.36)–(5.5.38). It is easy to do this by analogy with (5.5.38). Now, let

$$D_F(a_0, a_1, \ldots, a_n, x_1, \ldots, x_m) = 0$$

be a Diophantine representation of a function F. Let the equation

$$D_F(x_{m+1}, a_1, \ldots, a_n, x_1, \ldots, x_m) = 0$$

play the role of equation (5.4.1), constructing from it the machine M described in Section 5.4. When machine M halts, the tape will contain the tuple (5.4.9) with $y_m = F(a_1, \ldots, a_n)$. It suffices to apply the machine ERASE $2m + n + 2$ times and then the machine DELETE $k + 1$ times.

6. Let M be a Turing machine whose tape is infinite in both directions, and let $\alpha_1, \ldots, \alpha_w$ be the symbols that can be written on the tape. The operations of this machine can be simulated by another machine M' whose tape has only one end. The machine M' uses the alphabet

$$\star, \alpha_1, \ldots, \alpha_w, \beta_1, \ldots, \beta_w, \gamma_{11}, \ldots, \gamma_{ww}.$$

The tape of the machine M' imitates the tape of the machine M folded twice. To "strengthen" the tape of M', one has to make two copies of it. In one copy β_j should be replaced by empty cells and γ_{ij} replaced by α_j; in the other copy α_i should be replaced by empty cells and both β_i and γ_{ij} should be replaced by α_i. The cells with the marker "$\star$" should be cut off and the two semi-infinite tapes should be glued together to form the tape of the machine M.

The number of states doubles; i.e., for each state q_i of machine M there are two states $\mathrm{q}_i^{\mathrm{L}}$ and $\mathrm{q}_i^{\mathrm{R}}$ of machine M', corresponding to positioning the head of machine M on the left-hand side or on the right-hand side of the tape. The instructions for the machine M' arise in a natural way from those of the machine M. Namely, to each instruction $\mathrm{q}_i\alpha_j \Longrightarrow \alpha_k\mathrm{D}\mathrm{q}_l$ of the machine M, the machine M' has the counterparts

$$\begin{aligned}
\mathrm{q}_i^{\mathrm{R}}\alpha_j &\Longrightarrow \alpha_k\mathrm{D}\mathrm{q}_l^{\mathrm{R}} \\
\mathrm{q}_i^{\mathrm{R}}\gamma_{mj} &\Longrightarrow \gamma_{mk}\mathrm{D}\mathrm{q}_l^{\mathrm{R}} \\
\mathrm{q}_i^{\mathrm{L}}\beta_j &\Longrightarrow \beta_k\overline{\mathrm{D}}\mathrm{q}_l^{\mathrm{L}} \\
\mathrm{q}_i^{\mathrm{L}}\gamma_{jm} &\Longrightarrow \gamma_{km}\overline{\mathrm{D}}\mathrm{q}_l^{\mathrm{L}},
\end{aligned}$$

where $\overline{\mathrm{D}}$ is the direction opposite to D, i.e., $\overline{\mathrm{L}} = \mathrm{R}$, $\overline{\mathrm{R}} = \mathrm{L}$, $\overline{\mathrm{S}} = \mathrm{S}$. Similarly, to each instruction $\mathrm{q}_i\Lambda \Longrightarrow \alpha_k\mathrm{D}\mathrm{q}_l$ there are counterparts

$$\begin{aligned}
\mathrm{q}_i^{\mathrm{R}}\beta_m &\Longrightarrow \gamma_{mk}\mathrm{D}\mathrm{q}_l^{\mathrm{R}} \\
\mathrm{q}_i^{\mathrm{L}}\alpha_m &\Longrightarrow \gamma_{km}\overline{\mathrm{D}}\mathrm{q}_l^{\mathrm{L}} \\
\mathrm{q}_i^{\mathrm{R}}\Lambda &\Longrightarrow \alpha_k\mathrm{D}\mathrm{q}_l^{\mathrm{R}} \\
\mathrm{q}_i^{\mathrm{L}}\Lambda &\Longrightarrow \beta_k\overline{\mathrm{D}}\mathrm{q}_l^{\mathrm{L}}.
\end{aligned}$$

In addition, there are instructions

$$\begin{aligned}
\mathrm{q}_i^{\mathrm{R}}\star &\Longrightarrow \star\mathrm{R}\mathrm{q}_i^{\mathrm{L}} \\
\mathrm{q}_i^{\mathrm{L}}\star &\Longrightarrow \star\mathrm{R}\mathrm{q}_i^{\mathrm{R}}.
\end{aligned}$$

7. (a) The equivalence of register machines to Turing machines was established in the papers where the notion was introduced (see Commentary to Chapter 5).
(b) The papers Jones and Matiyasevich [1984, 1991] and Matiyasevich [1984] show how to use Diophantine equations (actually exponential Diophantine equations) to simulate register machines.

8. This was shown by Chudnovsky [1971, 1984]. Let

$$b \in \{\, 2^a \mid a \in \mathfrak{D} \,\} \iff \exists x_1 \dots x_m \, [D(b, x_1, \dots, x_m)];$$

then a representation of the set $\mathfrak{D}$ as required differs from

$$D^2(x_0^a, x_1, \dots, x_m) + (x_0 - 2)^2 = 0$$

only in a manner of writing.

6 Bounded Universal Quantifiers

1. This was just the original construction by Davis, Putnam, and Robinson [1961]. See Exercise 3.3 for a counterpart of (2.13).

2. Such a version of Dirichlet's Principle was used by Matiyasevich [1972a, 1973].

3. This can easily be done by applying the multiplicative version of Dirichlet's Principle from Exercise 6.2 m times (see, for example, Matiyasevich [1972a, 1973]).

4. See Matiyasevich [1972a, 1973].

5. This idea was suggested by Hirose and Iida [1973].

6. See Davis, Matiyasevich, and Robinson [1976].

7. One way to proceed would be to begin by finding an example of a Diophantine function that for given values of p and q will yield a number of equation (4.6.8) resulting from equation (4.6.4) by substituting the particular values of the parameters. We could then find a proof similar to the one given in Section 4.6 for the fact that the set $\mathfrak{H}_0$ is not Diophantine. However, it is easier instead of equation (4.6.8) to consider the equation in $\mathbf{m}+2$ unknowns

$$U_1^2(x_{\mathbf{m}+1}, x_{\mathbf{m}+2}, x_1, \dots, x_{\mathbf{m}}) + (x_{\mathbf{m}+1} - p)^2 + (x_{\mathbf{m}+2} - q)^2 = 0.$$

Let i and j be numbers such that

$$U_1(x_{\mathbf{m}+1}, x_{\mathbf{m}+2}, x_1, \ldots, x_{\mathbf{m}}) = P_i(x_1, \ldots, x_{\mathbf{m}+2}) - P_j(x_1, \ldots, x_{\mathbf{m}+2});$$

then the preceding equation can be rewritten as

$$P_i^2(x_1, \ldots, x_{\mathbf{m}+2}) + P_j^2(x_1, \ldots, x_{\mathbf{m}+2}) + x_{\mathbf{m}+1}^2 + x_{\mathbf{m}+2}^2 + p^2 + q^2$$
$$= 2(P_i(x_1, \ldots, x_{\mathbf{m}+2})P_j(x_1, \ldots, x_{\mathbf{m}+2}) + px_{\mathbf{m}+1} + qx_{\mathbf{m}+2}).$$

It easy to see that the number of that equation is the value of a polynomial $L(i, j, p, q)$ with rational coefficients.

8. This was proved by R. M. Robinson [1956]; see also Davis [1958].

7 Decision Problems in Number Theory

1. This was proved in Smoryński [1977] using singlefold exponential Diophantine representations.

2. See Smorinski [1977].

3. $d^c = \text{rem}(b^{c(c+1)(d+1)}, b^{(c+1)(d+1)} - d)$

4. Instead of (1.4.5), one could use the equation

$$(D_1^2(a_1, \ldots, a_n, x_1, \ldots, x_{m_1}) + y_1^2 + \cdots + y_{m_2}^2)$$
$$\cdot (D_2^2(a_1, \ldots, a_n, y_1, \ldots, y_{m_2}) + x_1^2 + \cdots + x_{m_1}^2) = 0.$$

5. This was proved in Matiyasevich [1973].

6. Originally this was established by Denef [1975]. Later, stronger results were obtained by Denef and Lipshitz [1978] and Denef [1980].

7. This was proved by Adler [1971]. Smoryński [1977] presented a simpler construction using the technique presented in Section 7.4.

8 Diophantine Complexity

1. This relation is defined by

$$\exists x \left[(b^2 + c^2 + 1)(d - e) = \left(\frac{b - c}{a + 1} \right)^2 + x + 1 \right].$$

2. Matiyasevich and Robinson [1975] showed that the following provides such a representation:

$$\exists x \left[\prod (x + \sum_{i=1}^{k} \pm \sqrt{S_i(a_1, \ldots, a_n)} W^{i-1}) = 0 \right].$$

Here W is an abbreviation for the sum $1 + S_1^2(a_1, \ldots, a_n) + \cdots + S_k^2(a_1, \ldots, a_n)$, and the product is taken over all 2^k ways of choosing the signs $\pm$, so that after elementary transformations all square roots disappear.

3. We have

$$\deg_p(a^2 q) = 2 \deg_p(a) + \deg_p(q) \equiv 1 \not\equiv 0 \equiv \deg_p(b^2) \pmod 2,$$

so $2kl \le \deg_p(a^2 q + b^2) = \min(\deg_p(a^2 q), \deg_p(b^2))$. It is easy to deduce by induction that the divisibility condition

$$q^{2^{m-1}} \mid a_1^2 q + \left(a_2^2 q + (\ldots (a_{m-1}^2 q + a_m^2)^2 \ldots)^2 \right)^2$$

implies the divisibility conditions $p \mid a_1, \ldots, p \mid a_m$.

4. Without loss of generality, we can assume that $B(a, b, w) \ge 4$, and hence

$$q_y \equiv -1 \pmod 4.$$

Thus, among the prime factors of q_y of the form $4n - 1$, there is at least one that divides it to an odd power.

5. This was originally proved by Matiyasevich [1972b]; another proof is given in Jones [1981].

6. See Matiyasevich and Robinson [1974].

7. See Jones [1981].

8. See Matiyasevich [1972b].

9. The set of all values of the polynomial $x_1^k + 2x_2^k + \cdots + 2^{m-1}x_m^k$ has this property provided that k is large enough relative to m. This example was discovered by R. M. Robinson and is presented in Matiyasevich and Robinson [1975].

10. This was accomplished by Jones and Matiyasevich [1982b] by using, instead of the universal representation (3.4.2) of a general binomial coefficient, the special representation of the central binomial coefficient from Exercise 3.4.

11. This result was obtained by Davis [1973b] as a corollary to the well-known Blum speed-up theorem [1967].

9 Decision Problems in Calculus

1. Assume that χ is the unique zero of an algebraic equation $A(\chi) = 0$ in an interval $[\alpha, \beta]$ with rational end points and that $A(\alpha) < 0,\ A(\beta) > 0$. The required Diophantine real can be defined as

$$\alpha < \frac{a-b}{c+1} < \frac{a-b+1}{c+1} < \beta \,\&\, c > d \,\&\, A\left(\frac{a-b}{c+1}\right) < 0 \,\&\, A\left(\frac{a-b+1}{c+1}\right) > 0.$$

2. Clearly, it suffices to show that there are Diophantine functions D, E, and F such that

$$\frac{D(n) - E(n)}{F(n) + 1} = \sum_{k=0}^{n} \frac{A(k) - B(k)}{C(k) + 1}.$$

Consider the tuples $\langle D(0), \ldots, D(n)\rangle$, $\langle E(0), \ldots, E(n)\rangle$, $\langle F(0), \ldots, F(n)\rangle$. The elements of these tuples are related to one another in a manner that can readily be expressed using bounded universal quantifiers.

3.
$$e = \frac{1}{0!} + \frac{1}{1!} + \cdots,$$

$$\frac{\pi}{4} = 1 - \frac{1}{3} + \frac{1}{5} - \cdots$$

Now it suffices to use Exercise 9.2.

4. Instead of the system $\Xi_1(\tau), \ldots, \Xi_k(\tau)$, one can work with the single function $\Xi(\sigma_1, \ldots, \sigma_k, \tau)$ satisfying the equations

$$\frac{\partial^2 \Xi}{\partial \sigma_i \partial \sigma_j} = 0$$

and replace Ξ_i by $\partial \Xi / \partial \sigma_i$ in (9.3.13) and (9.3.14). This idea was suggested by Adler [1969a].

10 Other Applications of Diophantine Representations

1. (a) It is the player B who is interested in the solvability of the first equation, so we replace the unknowns $x_1, \ldots, x_{\mathbf{m}}$ by $b_2, \ldots, b_{\mathbf{m}+1}$. Similarly, player A is interested in the solvability of the second equation, so we replace $y_1, \ldots, y_{\mathbf{m}}$ by $a_3, \ldots, a_{\mathbf{m}+2}$. By choosing the value of the last variable $b_{\mathbf{m}+2}$, player B will have the possibility of demonstrating that $a_3, \ldots, a_{\mathbf{m}+2}$ do not provide a solution. Finally, we obtain the Diophantine game defined by the equation

$$U_1(a_2, a_1, b_2, \ldots, b_{\mathbf{m}+1}) \cdot [U_1^2(b_1, a_1, a_3, \ldots, a_{\mathbf{m}+2}) - b_{\mathbf{m}+2} - 1] = 0.$$

(b) Having lost a game once, player B would always win later by choosing on the second move the number that had permitted player A to win on the third move.
(c) See Jones [1982a].

2. This was established by Ibarra and Leininger [1983] by showing that such programs are capable of computing the values of polynomials with natural number coefficients and of comparing numbers. See Shiryaev [1989] for improvements of this result.

Bibliography

Items are listed under the form of the author's name that is most common or, in the case of Russian authors, under the standard transliteration that we have used in this book. Whenever the name appears in a different form on a specific item, that form is indicated at the beginning of the entry for that item.

Zofia Adamowicz

1992 A sharp version of the bounded Matijasevich conjecture and the end-extension problem. *Journal of Symbolic Logic*, 57(2):597–616.

Leonard Adleman and Kenneth Manders

1976 Diophantine complexity. In 17th *Annual Symposium on Foundations of Computer Science*, pages 81–88, Houston, Texas, 25–26 October. IEEE.

Andrew Adler

1969a Some recursively unsolvable problems in analysis. *Proceedings of the American Mathematical Society*, 22(2):523–526.

1969b Extensions of non-standard models of number theory. *Zeitschrift für Mathematische Logik und Grundlagen der Mathematik*, 15(4):289–290.

1969c Existential formulas in arithmetic. *Dissertation Abstracts*, 29B(8):2962–2963.

1971 A. Adler. A reduction of homogeneous Diophantine problems. *The Journal of the London Mathematical Society (Second Series)*, 3(3):446–448.

S. V. Aĭvazyan

1983 O nekotorykh beskonechnykh tsepyakh mnogoobraziĭ grupp. *Doklady Akademiya Nauk Armyanskoĭ SSSR*, 26(5):198–200 (Russian).

1988 Ob odnoĭ probleme A. I. Mal'tseva. *Sibirskiĭ Matematicheskiĭ Zhurnal*, 29(6):3–11 (Russian). (Translated as S. V. Aivazyan. A problem of A. I. Mal'tsev. *Siberian Mathematical Journal*, 29(6):877–884, 1989.)

Yvette Amice

1964 Interpolation p-adique. *Bulletin de la Société Mathématique de France*, 92(2):117–180 (French).

George E. Andrews

1986 EΥΡΗΚΑ! num $= \Delta + \Delta + \Delta$. *Journal of Number Theory*, 23(3):285–293.

David J. Anick

1985 Diophantine equations, Hilbert series, and undecidable spaces. *Annals of Mathematics, Second Series*, 122(1):87–112.

Jean-Pierre Azra

1971 Relations Diophantiennes et la solution négative du 10e problème de Hilbert. In *Séminaire Bourbaki—vol. 1970/71, Exposés 382–399*, volume 244 of *Lecture Notes in Mathematics*, pages 11–28. Springer-Verlag (French).

A. Baker

1968 Contributions to the theory of Diophantine equations I. On the representation of integers by binary forms. *Philosophical Transactions of the Royal Society of London. Series A*, 263(1139):173–191.

D. Barton and J. P. Fitch

1972a Applications of algebraic manipulation programs in physics. *Reports on Progress in Physics*, 35(3):235–314.

1972b A review of algebraic manipulative programs and their application. *Computer Journal*, 15(4):362–381.

Friedrich L. Bauer

1988 For all primes greater than 3, $\binom{2p-1}{p-1} \equiv 1 \pmod{p^3}$ holds. *The Mathematical Intelligencer*, 10(3):42.

Walter Baur

1974 Über rekursive Strukturen. *Inventiones Mathematicae*, 23(2):89–95 (German).

Christoph Baxa

1993 A note on Diophantine representations. *The American Mathematical Monthly*, 100(2):138–143.

Lewis D. Baxter

1978 The undecidability of the third order dyadic unification problem. *Information and Control*, 38(2):170–178.

H. Beck

1975 Zur Entscheidbarkeit der Funktionalen Äquivalenz. In H. Brakhage, editor, *Automata Theory and Formal Languages: 2nd GI Conference*, volume 33 of *Lecture Notes in Computer Science*, pages 127–133. Springer-Verlag, May (German).

J. L. Bell and M. Machover

1977 *A Course in Mathematical Logic.* North Holland, Amsterdam.

Lenore Blum, Mike Shub, and Steve Smale

1989 On a theory of computation and complexity over the real numbers; *NP* completeness, recursive functions and universal machines. *Bulletin of the American Mathematical Society (New Series)*, 21(1):1–46.

Manuel Blum

1967 A machine-independent theory of the complexity of recursive functions. *Journal of the ACM*, 14(2):322–336.

L. A. Bokut'

1972 Nerazreshimost' problemy ravenstva i podalgebry konechno opredelennykh algebr Li. *Izvestiya Akademii Nauk SSSR. Seriya Matematichekaya*, 36(6):1173–1219 (Russian). (Translated as L. A. Bokut'. Unsolvability of the equality problem and subalgebras of finitely presented Lie algebras. *Mathematics of the USSR. Izvestiya*, 6(6):1153–1199, 1972.)

Dorothy Bollman and Miguel Laplaza

1978 Some decision problems for polynomial mappings. *Theoretical Computer Science*, 6(3):317–325.

E. Börger

1989 *Computability, Complexity, Logic.* North-Holland, Amsterdam.

J. L. Britton

1979 Integer solutions of systems of quadratic equations. *Mathematical Proceedings of the Cambridge Philosophical Society*, 86(3):385–389.

Manfred Broy, Martin Wirsing, and Peter Pepper

1987 On the algebraic definition of programming languages. *ACM Transactions on Programming Languages and Systems*, 9(1):54–99.

Cristian Calude and Gheorghe Păun

1983 Independent instances for some undecidable problems. *RAIRO Informatique Théorique*, 17(1):49–54.

D. Cantone, V. Cutello, and A. Policrito

1990 Set-theoretic reductions of Hilbert's Tenth Problem. In E. Börger, H. Kleine Büning, and M. M. Richter, editors, *CSL '89: 3rd Workshop on Computer Science Logic*, volume 440 of *Lecture Notes in Computer Science*, pages 65–75. Springer-Verlag.

Hans Georg Carstens

1977 The theorem of Matijasevich is provable in Peano's arithmetic by finitely many axioms. *Logique et Analyse*, 20:116–121.

B. F. Caviness

1970 On canonical forms and simplification. *Journal of the ACM*, 17(2):385–396.

Patrick Cegielski

1990 La théorie de corps réel-clos inductifs est une extension conservative de l'Arithmétique de Peano. *Comptes Rendus de l'Académie des Sciences. Série I. Mathématique*, 310(5):239–242 (French).

Gregory J. Chaitin

1974 Information-theoretic limitations of formal systems. *Journal of the ACM*, 21(3):403–424.

1987 *Algorithmic Information Theory*, volume 1 of *Cambridge Tracts in Theoretical Computer Science*. Cambridge University Press, Cambridge.

Tat-hung Chan

1987 On two-way weak counter machines. *Mathematical Systems Theory*, 20(1):31–41.

S. Chowla

1965 *The Riemann Hypothesis and Hilbert's Tenth Problem*. Gordon and Breach, New York.

Gregory V. Chudnovsky

1970 G. V. Chudnovskiĭ. Diofantovy predikaty. *Uspekhi Matematicheskikh Nauk*, 25(4):185–186 (Russian).

1971 G. V. Chudnovskiĭ. Nekotorye arifmeticheskie problemy. Preprint IM-71-3, Akademiya Nauk Ukrainskoĭ SSR. Institut Matematiki, Kiev.

1984 *Contributions to the Theory of Transcendental Numbers*. Number 19 in Mathematical Surveys and Monographs. American Mathematical Society, Providence.

Alonzo Church

1936 An unsolvable problem of elementary number theory. *American Journal of Mathematics*, 58:345–363.

Lori A. Clarke and Debra J. Richardson

1985 Applications of symbolic evaluation. *Journal of Systems and Software*, 5(1):15–35.

Jacques Cohen

1990 Constraint logic programming languages. *Communications of the ACM*, 33(7):52–68.

J. H. Conway and A. J. Jones

1976 Trigonometric Diophantine equations (On vanishing sums of roots of unity). *Acta Arithmetica*, 30(3):229–240.

Wiktor Dańko

1980 A criterion of undecidability of algorithmic theories. In P. Dembiński, editor, *Mathematical Foundations of Computer Science 1980*, volume 88 of *Lecture Notes in Computer Science*, pages 205–218, Rydzyna, Poland. Springer-Verlag.

1981 A criterion of undecidability of algorithmic theories. *Fundamenta Informaticae*, IV(3):605–628.

Martin Davis

1950 Arithmetical problems and recursively enumerable predicates (abstract). *Journal of Symbolic Logic*, 15(1):77–78.

1953 Arithmetical problems and recursively enumerable predicates. *Journal of Symbolic Logic*, 18(1):33–41.

1958 *Computability and Unsolvability*. McGraw Hill, New York. Reprint. Dover Publications, New York, 1982.

Martin Davis (continued)

1962 Applications of recursive function theory to number theory. In *Recursive Function Theory*, volume 5 of *Proceedings of Symposia in Pure Mathematics*, pages 135–138, Providence, Rhode Island. American Mathematical Society.

1963 Extensions and corollaries of recent work on Hilbert's Tenth Problem. *Illinois Journal of Mathematics*, 7(2):246–250.

1966 Diophantine equations and recursively enumerable sets. In E. R. Caianiello, editor, *Automata Theory*, pages 146–152. Academic Press, New York.

1968 One equation to rule them all. *Transactions of the New York Academy of Sciences. Series II*, 30(6):766–773.

1971 An explicit Diophantine definition of the exponential function. *Communications on Pure and Applied Mathematics*, 24(2):137–145.

1972 On the number of solutions of Diophantine equations. *Proceedings of the American Mathematical Society*, 35(2):552–554.

1973a Hilbert's Tenth Problem is unsolvable. *The American Mathematical Monthly*, 80(3):233–269. Reprinted with corrections in the Dover edition of *Computability and Unsolvability* (Davis [1958]).

1973b Speed-up theorems and Diophantine equations. In Randall Rustin, editor, *Courant Computer Science Symposium 7: Computational Complexity*, pages 87–95. Algorithmics Press, New York.

1977 Unsolvable problems. In Jon Barwise, editor, *Handbook of Mathematical Logic*, volume 90 of *Studies in Logic and the Foundations of Mathematics*, chapter C.2, pages 567–594. North Holland, Amsterdam.

1989 Teaching the incompleteness theorem. In R. Ferro, C. Bonotto, S. Valentini, and A. Zanardo, editors, *Logic Colloquium '88*, volume 127 of *Studies in Logic and the Foundations of Mathematics*, pages 385–392, Amsterdam. North-Holland.

Martin Davis, Yu. V. Matiyasevich, and Julia Robinson

1976 Martin Davis, Yuri Matijasevič, and Julia Robinson. Hilbert's Tenth Problem. Diophantine equations: positive aspects of a negative solution. In *Mathematical Developments Arising from Hilbert Problems*, volume 28 of *Proceedings of Symposia in Pure Mathematics*, pages 323–378, Providence, Rhode Island. American Mathematical Society.

Martin Davis and Hilary Putnam

1958 Reductions of Hilbert's Tenth Problem. *Journal of Symbolic Logic*, 23(2):183–187.

1959a M. D. Davis and Hilary Putnam. On Hilbert's Tenth Problem. *Notices of the American Mathematical Society*, 6(5):544.

Martin Davis and Hilary Putnam (continued)

1959b A computational proof procedure; Axioms for number theory; Research on Hilbert's Tenth Problem. O.S.R. Report AFOSR TR59-124, U.S. Air Force, October.

1963 Diophantine sets over polynomial rings. *Illinois Journal of Mathematics*, 7(2):251–256.

Martin Davis, Hilary Putnam, and Julia Robinson

1961 The decision problem for exponential Diophantine equations. *Annals of Mathematics, Second Series*, 74(3):425–436.

J. Denef

1975 Hilbert's Tenth Problem for quadratic rings. *Proceedings of the American Mathematical Society*, 48(1):214–220.

1978a Diophantine sets over $\mathbf{Z}[T]$. *Proceedings of the American Mathematical Society*, 69(1):148–150.

1978b The Diophantine problem for polynomial rings and fields of rational functions. *Transactions of the American Mathematical Society*, 242:391–399.

1980 Diophantine sets over algebraic integer rings. II. *Transactions of the American Mathematical Society*, 257(1):227–236.

J. Denef and L. Lipshitz

1978 Diophantine sets over some rings of algebraic integers. *The Journal of the London Mathematical Society (Second Series)*, 18(3):385–391.

1984 Power series solutions of algebraic differential equations. *Mathematische Annalen*, 267(2):213–238.

1989 Decision problems for differential equations. *Journal of Symbolic Logic*, 54(3):941–950.

Leonard Eugene Dickson

1939 *Modern Elementary Theory of Numbers*. The University of Chicago Press, Chicago.

Diophantus

1974 *Arifmetika i kniga o mnogougol'nykh chislakh*. Nauka, Moscow (Russian).

Michael L. Dowling

1990 Optimal code parallelization using unimodular transformations. *Parallel Computing*, 16(2–3):157–171.

V. G. Durnev

1974 Ob uravneniyakh na svobodnykh polugruppakh i gruppakh. *Matematicheskie Zametki*, 16(5):717–724 (Russian). (Translated as V. G. Durnev. On equations in free semigroups and groups. *Mathematical Notes*, 16(5–6):1024–1028, 1975.)

V. G. Durnev and N. B. Korablëva

1988 Ob odnoĭ algoritmicheski nerazreshimoĭ zadache dlya sistem lineĭnykh diofantovykh uravneniĭ. In *Voprosy teorii grupp i gomologich. algebry.*, pages 93–96. Yaroslavskiĭ Gosudarstvennyĭ Universitet, Yaroslavl' (Russian).

Verena H. Dyson, James P. Jones, and John C. Shepherdson

1982 Some Diophantine forms of Gödel's theorem. *Archiv für Mathematische Logik und Grundlagenforschung*, 22(1–2):51–60.

William M. Farmer

1991 A unification-theoretic method for investigating the k-provability problem. *Annals of Pure and Applied Logic*, 51(3):173–214.

Jens Erik Fenstad

1971 Hilberts 10. Problem. *Nordisk Matematisk Tidskrift*, 19(1–2):5–14 (Norwegian).

J. P. Fitch

1973 On algebraic simplification. *Computer Journal*, 16(1):23–27.

Ulf Friedrichsdorf

1976 Einige Bemerkungen zur Peano-Arithmetik. *Zeitschrift für Mathematische Logik und Grundlagen der Mathematik*, 22(5):431–436 (German).

G. Gagliardi and S. Tulipani

1990 On algebraic specifications of computable algebras with the discriminator technique. *RAIRO Informatique Théorique et Applications*, 24(5):429–440.

Haim Gaifman

1975 A note on models and submodels of arithmetic. In *Conference in Mathematical Logic—London '70*, volume 255 of *Lecture Notes in Mathematics*, pages 128–144. Springer-Verlag.

Haim Gaifman and Constantine Dimitracopoulos

1982 Fragments of Peano's arithmetic and the MRDP theorem. In *Logic and Algorithmic*, Monographie N° 30 de L'Enseignement Mathématique, pages 187–206. Université de Genève.

Michael R. Garey and David S. Johnson

1979 *Computers and Intractability: A Guide to the Theory of NP-Completeness.* W. H. Freeman, San Francisco.

Dainis Geimanis

1990 On possibilities of one-way synchronized and alternating automata. In B. Rovan, editor, *Mathematical Foundations of Computer Science 1990*, volume 452 of *Lecture Notes in Computer Science*, pages 292–299. Springer-Verlag.

Giorgio Germano

1976 An arithmetical reconstruction of the liar's antinomy using addition and multiplication. *Notre Dame Journal of Formal Logic*, 17(3):457–461.

Kurt Gödel

1931 Über formal unentscheidbare Sätze der Principia Mathematica und verwandter Systeme I. *Monatshefte für Mathematik und Physik*, 38(1):173–198 (German). Reprinted with translation in Solomon Feferman et al., editors, *Kurt Gödel. Collected Works*, volume I, pages 144–195. Oxford University Press, 1986. (Also translated as Kurt Gödel. On formally undecidable propositions in *Principia Mathematica* and related systems I. In Martin Davis, editor, *The Undecidable*, pages 4–38. Raven Press, Hewlett, New York, 1965. Translation also available in Jean van Heijenoort, editor, *From Frege to Gödel: A Source Book in Mathematical Logic, 1879–1831*, pages 596–616. Harvard University Press, Cambridge, Massachusetts, 1967.

Warren D. Goldfarb

1981 The undecidability of the second-order unification problem. *Theoretical Computer Science*, 13(2):225–230.

R. L. Goodstein

1975 Hilbert's Tenth Problem and the independence of recursive difference. *The Journal of the London Mathematical Society (Second Series)*, 10(2):175–176.

R. L. Goodstein and R. D. Lee

1966 A decidable class of equations in recursive arithmetic. *Zeitschrift für Mathematische Logik und Grundlagen der Mathematik*, 12:235–239.

I. L. Gorskiĭ and A. M. Yakubovich

1985 Voprosy organizatsii dialoga v sisteme imitatsionnogo modelirovaniya. *Programmirovanie*, no. 5:77–82 (Russian). (Translated as I. L. Gorskii and A. M. Yakubovich. Organization of the conversational mode in a simulation system. *Programming and Computer Software*, 11(5):310–314, 1986.)

D. Guaspari

1979 Partially conservative extensions of arithmetic. *Transactions of the American Mathematical Society*, 254:47–68.

Eitan M. Gurari

1985 Decidable problems for powerful programs. *Journal of the ACM*, 32(2):466–483.

Eitan M. Gurari and Oscar H. Ibarra

1979 An NP-complete number-theoretic problem. *Journal of the ACM*, 26(3):567–581.

1982 Two-way counter machines and Diophantine equations. *Journal of the ACM*, 29(3):863–873.

Michel Hack

1976 The equality problem for vector addition systems is undecidable. *Theoretical Computer Science*, 2(1):77–95.

Petr Hájek and Tomáš Havránek

1978 *Mechanizing Hypothesis Formation: Mathematical Foundations for a General Theory*. Springer-Verlag, Berlin.

David Harel, Amir Pnueli, and Jonathan Stavi

1983 Propositional dynamic logic of nonregular programs. *Journal of Computer and System Sciences*, 26(2):222–243.

G. Hasenjaeger

1984 Universal Turing machines (UTM) and Jones-Matiyasevich-masking. In *Logic and Machines: Decision Problems and Complexity (Münster, 1983)*, volume 171 of *Lecture Notes in Computer Science*, pages 248–253, Berlin. Springer-Verlag.

William S. Hatcher and Bernard R. Hodgson

1981 Complexity bounds on proofs. *Journal of Symbolic Logic*, 46(2):255–258.

Ivan Havel

1973 O desátém Hilbertově problému. *Pokroky Matematiky, Fyziky a Astronomie*, 18(4):185–192 (Czech).

Thomas L. Heath

1910 *Diophantus of Alexandria: A Study in the History of Greek Algebra.* Cambridge University Press, Cambridge, second edition. Reprint. Dover Publications, New York, 1964.

Jan Heering

1986 Partial evaluation and ω-completeness of algebraic specifications. *Theoretical Computer Science*, 43(2–3):149–167.

Hans Hermes

1972 Die Unlösbarkeit des zehnten Hilbertschen Problems. *L'Enseignement Mathématique* (2), 18(1):47–56 (German).

1978 Die Unlösbarkeit des zehnten Hilbertschen Problems. *Mathematische und Naturwissenschaftliche Unterricht*, 31(5):260–263 (German).

Oskar Herrman

1971 A non-trivial solution of the Diophantine equation $9(x^2+7y^2)^2-7(u^2+7v^2)^2=2$. In A. O. L. Atkin and B. J. Birch, editors, *Computers in Number Theory*, pages 207–212. Academic Press, London.

D. Hilbert

1900 Mathematatische Probleme. Vortrag, gehalten auf dem internationalen Mathematiker-Kongreß zu Paris 1900. *Nachrichten von der Königliche Gesellschaft der Wissenschaften zu Göttingen*, pages 253–297 (German). Reprinted in David Hilbert. *Gesammelte Abhandlungen*, pages 290–329. Verlag von Julius Springer, Berlin, 1935. (Translated as D. Hilbert. Mathematical problems. Lecture delivered before the International Congress of Mathematicians at Paris in 1900. *Bulletin of the American Mathematical Society*, 8:437–479, 1902. Translation reprinted in *Mathematical Developments Arising from Hilbert Problems*, volume 28 of *Proceedings of Symposia in Pure Mathematics*, pages 1–34. American Mathematical Society, Providence, Rhode Island, 1976.

Ken Hirose

1968 A conjecture on Hilbert's 10th Problem. *Commentarii Mathematici Universitatis Sancti Pauli*, 17(1):31–34.

1973 On Hilbert's Tenth Problem (negative solution). *Sûgaku*, 25(1):1–9 (Japanese).

Ken Hirose and Shigeaki Iida

1973 A proof of negative answer to Hilbert's 10th Problem. *Proceedings of the Japan Academy*, 49(1):10–12.

Joram Hirschfeld

1975 J. Hirschfeld. Models of arithmetic and recursive functions. *Israel Journal of Mathematics*, 20(2):111–126.

Joram Hirschfeld and William Wheeler

1975 *Forcing, Arithmetic, Division Rings*, volume 454 of *Lecture Notes in Mathematics*. Springer-Verlag.

Bernard R. Hodgson and Clement F. Kent

1983 A normal form for arithmetical representation of $\mathcal{NP}$-sets. *Journal of Computer and System Sciences*, 27(3):378–388.

V. E. Hoggatt, Jr. and Marjorie Bicknell-Johnson

1978 Divisibility by Fibonacci and Lucas squares. *The Fibonacci Quarterly*, 15(1):3–8.

Rodney R. Howell, Louis E. Rosier, Dung T. Huynh, and Hsu-Chun Yen

1986 Some complexity bounds for problems concerning finite and 2-dimensional vector addition systems with states. *Theoretical Computer Science*, 46(2–3):107–140.

Verena Huber-Dyson

1984 HNN-constructing finite groups. In A. C. Kim and B. H. Neumann, editors, *Groups—Korea 1983. Proceedings*, volume 1098 of *Lecture Notes in Mathematics*, pages 58–62. Springer-Verlag.

Gérard Huet and Derek C. Oppen

1980 Equations and rewrite rules: A survey. In Ronald V. Book, editor, *Formal Language Theory: Perspectives and Open Problems*, pages 349–405. Academic Press, New York.

Oscar H. Ibarra

1982 2DST mappings on languages and related problems. *Theoretical Computer Science*, 19(2):219–227.

Oscar H. Ibarra and Brian S. Leininger

1980 The complexity of the equivalence problem for straight-line programs. In *Conference Proceedings of the Twelfth Annual ACM Symposium on Theory of Computing*, pages 273–280, Los Angeles, California, 28-20 April.

1982 Straight-line programs with one input variable. *SIAM Journal on Computing*, 11(1):1–14.

1983 On the simplification and equivalence problems for straight-line programs. *Journal of the ACM*, 30(3):641–656.

R. G. Jeroslow

1973 There cannot be any algorithm for integer programming with quadratic constraits. *Operations Research*, 21(1):221–224.

James P. Jones

1974 Recursive undecidability—an exposition. *The American Mathematical Monthly*, 81(7):724–738.

1975 Diophantine representation of the Fibonacci numbers. *The Fibonacci Quarterly*, 13(1):84–88.

James P. Jones (continued)

1978 Three universal representations of recursively enumerable sets. *Journal of Symbolic Logic*, 43(2):335–351.

1979 Diophantine representation of Mersenne and Fermat primes. *Acta Arithmetica*, 35(3):209–221.

1980 Undecidable Diophantine equations. *Bulletin of the American Mathematical Society (New Series)*, 3(2):859–862.

1981 Classification of quantifier prefixes over Diophantine equations. *Zeitschrift für Mathematische Logik und Grundlagen der Mathematik*, 27(5):403–410.

1982a J. P. Jones. Some undecidable determined games. *International Journal of Game Theory*, 11(2):63–70.

1982b Universal Diophantine equation. *Journal of Symbolic Logic*, 47(3):549–571.

1991 Dzh. P. Dzhons. Vychislitel'naya slozhnost' vyigryvayushchikh strategiĭ v polinomial'nykh igrakh dvukh lits. *Zapiski Nauchnykh Seminarov Leningradskogo Otdeleniya Matematicheskogo Instituta im. V. A. Steklova AN SSSR (LOMI)*, 192:69–73 (Russian).

James P. Jones, H. Levitz, and A. J. Wilkie

1986 J. P. Jones, H. Levitz, and A. J. Wilkie. Classification of quantifier prefixes over exponential Diophantine equations. *Zeitschrift für Mathematische Logik und Grundlagen der Mathematik*, 32(5):399–406.

James P. Jones and Yu. V. Matiyasevich

1982a J. P. Jones and Ju V. Matijasevič. Exponential Diophantine representation of recursively enumerable sets. In J. Stern, editor, *Proceedings of the Herbrand Symposium: Logic Colloquium '81*, volume 107 of *Studies in Logic and the Foundations of Mathematics*, pages 159–177, Amsterdam. North Holland.

1982b J. P. Jones and Ju. V. Matijasevič. A new representation for the symmetric binomial coefficient and its applications. *Les Annales des Sciences Mathématiques du Québec*, 6(1):81–97.

1983 J. P. Jones and Yu. V. Matijasevich. Direct translation of register machines into exponential Diophantine equations. In L. Priese, editor, *Report on the 1st GTI-workshop*, number 13, pages 117–130, Reihe Theoretische Informatik, Universität-Gesamthochschule Paderborn.

1984 J. P. Jones and Y. V. Matijasevič. Register machine proof of the theorem on exponential Diophantine representation of enumerable sets. *Journal of Symbolic Logic*, 49(3):818–829.

1991 J. P. Jones and Y. V. Matijasevič. Proof of recursive unsolvability of Hilbert's Tenth Problem. *The American Mathematical Monthly*, 98(8):689–709.

James P. Jones, Daihachiro Sato, Hideo Wada, and Douglas Wiens

1976 Diophantine representation of the set of prime numbers. *The American Mathematical Monthly*, 83(6):449–464.

Deborah Joseph

1983 Polynomial time computations in models of ET. *Journal of Computer and System Sciences*, 26(3):311–338.

Deborah Joseph and Paul Young

1981 A survey of some recent results on computational complexity in weak theories of arithmetic. In J. Gruska and M. Chytil, editors, *Mathematical Foundations of Computer Science 1981*, volume 118 of *Lecture Notes in Computer Science*, pages 46–60. Springer-Verlag.

Irving Kaplansky

1977 Hilbert's problems. Lecture notes, Department of Mathematics, University of Chicago.

Juhani Karhumäki

1977 The decidability of the equivalence problem for polynomially bounded D0L sequences. *RAIRO Informatique Théorique*, 11(1):17–28.

Tadao Kasami and Nobuki Tokura

1971 Equivalence problem of programs without loops. *Systems-Computers-Controls*, 2(4):83–84.

Richard Kaye

1987 R. Kaye. Parameter free induction in arithmetic. *Seminarbericht. Humbold-Universiät. Berlin. Section Mathematik (DDR)*, pages 70–81.

1990 Diophantine induction. *Annals of Pure and Applied Logic*, 46(1):1–40.

1991 *Models of Peano Arithmetic*. Clarendon Press, Oxford.

Richard Kaye, J. Paris, and C. Dimitracopoulos

1988 R. Kaye, J. Paris, and C. Dimitracopoulos. On parameter free induction schemas. *Journal of Symbolic Logic*, 53(4):1082–1097.

Clement F. Kent and Bernard R. Hodgson

1982 An arithmetical characterization of NP. *Theoretical Computer Science*, 21(3):255–267.

O. G. Kharlampovich

1989a Problema ravenstva dlya razreshimykh algebr Li i grupp. *Matematicheskiĭ Sbornik*, 180(8):1033–1066 (Russian). (Translated as O. G. Kharlampovich. The word problem for solvable Lie algebras and groups. *Mathematics of the USSR. Sbornik*, 67(2):489–525, 1990.)

1989b Konechno opredelennye razreshimye gruppy i algebry Li s nerazreshimoĭ problemoĭ ravenstva. *Matematicheskie Zametki*, 46(3):80–92 (Russian). (Translated as O. G. Kharlampovich. Finitely presented solvable groups and Lie algebras with unsolvable equality problem. *Mathematical Notes*, 46(3):731–739, 1990.)

K. H. Kim and F. W. Roush

1989 Problems equivalent to rational Diophantine solvability. *Journal of Algebra*, 124(2):493–505.

Péter Kiss

1979 Diophantine representation of generalized Fibonacci numbers. *Elemente der Mathematik*, 34(6):129–132.

Yu. G. Kleĭman

1979 Tozhdestva i nekotorye algoritmicheskie problemy v gruppakh. *Doklady Akademii Nauk SSSR*, 244(4):814–818 (Russian). (Translated as Ju. G. Kleĭman. Identities and some algorithmic problems in groups. *Soviet Mathematics. Doklady*, 20(1):115–119, 1979.)

A. N. Kolmogorov and V. A. Uspenskiĭ

1958 K opredeleniyu algoritma. *Uspekhi Matematicheskikh Nauk*, 13(4):3–28 (Russian). (Translated as A. N. Kolmogorov and V. A. Uspenskiĭ. On the definition of an algorithm. *American Mathematical Society Translations. Series 2*, 29:217–245, 1963.)

Moshe Koppel

1979 Some decidable Diophantine problems: positive solution to a problem of Davis, Matijasevič and Robinson. *Proceedings of the American Mathematical Society*, 77(3):319–323.

N. K. Kosovskiĭ

1967 Dostatochnye usloviya nepolnoty dlya formalizatsiĭ chasteĭ arifmetiki. *Zapiski Nauchnykh Seminarov Leningradskogo Otdeleniya Matematicheskogo Instituta im. V. A. Steklova AN SSSR (LOMI)*, 4:44–57 (Russian). (Translated as N. K. Kosovskii. Sufficient conditions of incompleteness for the formalization of parts of arithmetic. *Seminars in Mathematics, V. A. Steklov Mathematical Institute*, 4:15–20, 1969.)

1971 O Diofantovykh predstavleniyakh posledovatel'nosti resheniĭ uravneniya Pellya. *Zapiski Nauchnykh Seminarov Leningradskogo Otdeleniya Matematicheskogo Instituta im. V. A. Steklova AN SSSR (LOMI)*, 20:49–59 (Russian). (Translated as N. K. Kosovskii. Diophantine representation of the sequence of solutions of the Pell equation. *Journal of Soviet Mathematics*, 1(1):28–35, 1973.)

1974 O reshenii sistem, sostoyashchikh odnovremenno iz uravneniĭ b slovakh i neravenstv v dlinakh v slovakh. *Zapiski Nauchnykh Seminarov Leningradskogo Otdeleniya Matematicheskogo Instituta im. V. A. Steklova AN SSSR (LOMI)*, 40:24–29 (Russian). (Translated as N. K. Kosovskii. Solutions of systems consisting of word equations and inequalities in lengths of words. *Journal of Soviet Mathematics*, 8(3):262–265, 1977.)

G. Kreisel

1958 Mathematical significance of consistency proofs. *Journal of Symbolic Logic*, 23(2):155–182.

1962 A3061: Davis, Martin; Putnam, Hilary; Robinson, Julia. The decision problem for exponential Diophantine equations. *Mathematical Reviews*, 24A(6A):573.

V. Yu. Kryauchyukas

1979 Diofantovo predstavlenie sovershennykh chisel. *Zapiski Nauchnykh Seminarov Leningradskogo Otdeleniya Matematicheskogo Instituta im. V. A. Steklova AN SSSR (LOMI)*, 88:78–89 (Russian). (Translated as V. Yu. Kryauchyukas. Diophantine representation of perfect numbers. *Journal of Soviet Mathematics*, 20(4):2307–2313, 1982.)

G. A. Kucherov

1988 A new quasi-reducibility testing algorithm and its application to proofs by induction. In J. Grabowski, P. Lescanne, and W. Wechler, editors, *Algebraic and Logic Programming*, volume 343 of *Lecture Notes in Computer Science*, pages 204–213. Springer-Verlag.

E. E. Kummer

1852 Über die Ergänzungssätze zu den allgemeinen Reciprocitätsgesetzen. *Journal für die Reine und Angewandte Mathematik*, 44:93–146 (German). Reprinted in André Weil, editor, *Ernst Eduard Kummer. Collected Papers*, volume 1, pages 485–538. Springer-Verlag, Berlin, 1975.

J. L. Lagrange

1772 Démonstration d'un théorème d'arithmétique. *Nouveaux Mémoires de l'Académie royale des Sciences et Belles-Lettres de Berlin, année 1770*, pages 123–133 (French).

Joachim Lambek

1961 How to program an infinite abacus. *Canadian Mathematical Bulletin*, 4:295–302.

Janusz Laski

1986 An algorithm for the derivation of codefinitions in computer programs. *Information Processing Letters*, 23(2):85–90.

H. W. Lenstra, Jr.

1982 Primality testing. In H. W. Lenstra, Jr. and R. Tijdeman, editors, *Computational Methods in Number Theory. Part 1*, number 154 in Mathematical Centre Tracts, pages 55–77. Mathematisch Centrum, Amsterdam.

Hilbert Levitz

1985 Decidability of some problem pertaining to base 2 exponential Diophantine equations. *Zeitschrift für Mathematische Logik und Grundlagen der Mathematik*, 31(2):109–115.

John S. Lew and Arnold L. Rosenberg

1975 Polynomial mappings from $N \times N$ into N. Technical Report RC-5761, IBM. A revised and expanded version of this paper appeared in the *Journal of Number Theory* (Lew and Rosenberg [1978a, 1978b]).

1978a Polynomial indexing of integer lattice-points. I. General concepts and quadratic polynomials. *Journal of Number Theory*, 10(2):192–214.

1978b Polynomial indexing of integer lattice-points. II. Nonexistence results for higher-degree polynomials. *Journal of Number Theory*, 10(2):215–243.

L. Lipshitz

1977 Leonard Lipshitz. Undecidable existential problems for addition and divisibility in algebraic number rings. II. *Proceedings of the American Mathematical Society*, 64(1):122–128.

1978a The Diophantine problem for addition and divisibility. *Transactions of the American Mathematical Society*, 235:271–283.

1978b Undecidable existential problems for addition and divisibility in algebraic number rings. *Transactions of the American Mathematical Society*, 241:121–128.

1979 Diophantine correct models of arithmetic. *Proceedings of the American Mathematical Society*, 73(1):107–108.

M. Livesey, J. Siekmann, P. Szabó, and E. Unvericht

1979 Unification problems for combinations of associativity, commutativity, distributivity and idempotence axioms. In William H. Joyner, Jr., editor, *Proceedings of the Fourth Workshop on Automated Deduction*, pages 175–184, Austin, Texas, 1–3 February.

Roger C. Lyndon and Paul E. Schupp

1977 *Combinatorial Group Theory*. Springer-Verlag, Berlin.

A. Macintyre and H. Simmons

1973 Gödel's diagonalization technique and related properties of theories. *Colloquium Mathematicum*, 28(2):165–180.

A. I. Mal'tsev

1968 O nekotorykh pogranichnykh voprosakh algebry i logiki. In *Trudy Mezhdunarodnogo Kongressa Matematikov (Moskva, 1966)*, pages 217–231. Mir (Russian).

Miroslav Maňas

1971 Nekonvexní úlohy matematického programování. *Ekonomicko-Matematický Obzor*, 7(4):390–405 (Czech).

Kenneth L. Manders

1980 Computational complexity of decision problems in elementary number theory. In L. Pacholski, J. Wierzejewski, and A. J. Wilkie, editors, *Model Theory of Algebra and Arithmetic*, volume 834 of *Lecture Notes in Mathematics*, pages 211–227. Springer-Verlag.

Kenneth L. Manders and Leonard Adleman

1978 *NP*-complete decision problems for binary quadratics. *Journal of Computer and System Sciences*, 16(2):168–184.

Larry Michael Manevitz

1976a Robinson forcing is not absolute. *Israel Journal of Mathematics*, 25:211–232.

1976b Internal end-extensions of Peano arithmetic and a problem of Gaifman. *The Journal of the London Mathematical Society (Second Series)*, 13(1):80–82.

Yu. I. Manin

1973 Desyataya problema Gil'berta. *Sovremennye problemy matematiki,* 1:5–37. Akademiya Nauk SSSR. Vsesoyuznyĭ Institut Nauchnoĭ i Tekhnicheskoĭ Informatsii, Moscow (Russian). (Translated as Yu. I. Manin. Hilbert's Tenth Problem. *Journal of Soviet Mathematics*, 3(2):161–184, 1975.)

1977 *A Course in Mathematical Logic.* Springer-Verlag, New York.

Yu. I. Manin and A. A. Panchishkin

1990 Vvedenie v teoriyu chisel. *Sovremennye problemy matematiki, Fundamental'ny napravlenia*, 49:5–348. Itogi nauki i tekhniki (Russian).

M. Margenstern

1981 Le théorèm de Matiyassévitch et résultats connexes. In C. Berline, K. McAloon, and J.-P. Ressayre, editors, *Model Theory and Arithmetic*, volume 890 of *Lecture Notes in Mathematics*, pages 198–241. Springer-Verlag (French).

Per Martin-Löf

1970 *Notes on Constructive Mathematics.* Almqvist and Wiksell, Stockholm.

V. I. Mart'yanov

1977 Universal'nye rasshirennye teorii tselykh chisel. *Algebra i Logika*, 16(5):588–602 (Russian). (Translated as V. I. Mart'yanov. Extended universal theories of the integers. *Algebra and Logic*, 16(5):395–405, 1978.)

S. Yu. Maslov

1967 Ponyatie strogoĭ predstavimosti v obshcheĭ teorii ischisleniĭ. *Trudy Matematicheskogo Instituta im. V. A. Steklova AN SSSR*, 93:3–42 (Russian). (Translated as S. Yu. Maslov. The concept of strict representability in the general theory of calculi. *Proceedings of Steklov Institute of Mathematics*, 93:1–50, 1970.)

Atilla Máté

1990 Nondeterministic polynomial-time computations and models of arithmetic. *Journal of the ACM*, 37(1):175–193.

Yu. V. Matiyasevich

1968a Svyaz' sistem uravneniĭ v slovakh i dlinakh s 10-ĭ problemoĭ Gil'berta. *Zapiski Nauchnykh Seminarov Leningradskogo Otdeleniya Matematicheskogo Instituta im. V. A. Steklova AN SSSR (LOMI)*, 8:132–144 (Russian). (Translated as Yu. V. Matiyasevich. The connection between Hilbert's Tenth Problem and systems of equations between words and lengths. *Seminars in Mathematics, V. A. Steklov Mathematical Institute*, 8:61–67, 1970.)

1968b Dve reduktsii 10-ĭ problemy Gil'berta. *Zapiski Nauchnykh Seminarov Leningradskogo Otdeleniya Matematicheskogo Instituta im. V. A. Steklova AN SSSR (LOMI)*, 8:145–158 (Russian). (Translated as Yu. V. Matiyasevich. Two reductions of Hilbert's Tenth Problem. *Seminars in Mathematics, V. A. Steklov Mathematical Institute*, 8:68–74, 1970.)

1968c Arifmeticheskie predstavleniya stepeneĭ. *Zapiski Nauchnykh Seminarov Leningradskogo Otdeleniya Matematicheskogo Instituta im. V. A. Steklova AN SSSR (LOMI)*, 8:159–165 (Russian). (Translated as Yu. V. Matiyasevich. Arithmetic representations of powers. *Seminars in Mathematics, V. A. Steklov Mathematical Institute*, 8:75–78, 1970.)

1970 Diofantovost' perechislimykh mnozhestv. *Doklady Akademii Nauk SSSR*, 191(2):279–282 (Russian). (Translated as Ju. V. Matijasevič. Enumerable sets are Diophantine. *Soviet Mathematics. Doklady*, 11(2):354–358, 1970.)

1971a Diofantovo predstavlenie mnozhestva prostykh chisel. *Doklady Akademii Nauk SSSR*, 196(4):770–773 (Russian). (Translated as Ju. V. Matijasevič. Diophantine representation of the set of prime numbers. *Soviet Mathematics. Doklady*, 12(1):249–254, 1971.)

1971b Diofantovo predstavlenie perechislimykh predikatov. *Izvestiya Akademii Nauk SSSR. Seriya Matematichekaya*, 35(1):3–30 (Russian). (Translated as Ju. V. Matijasevič. Diophantine representations of enumerable predicates. *Mathematics of the USSR. Izvestiya*, 15(1):1–28, 1971.)

1971c Yu. V. Matijasevič. Diophantine representation of recursively enumerable predicates. In *Actes du Congrès International des Mathématiciens (Nice* 1970), volume 1, pages 235–238, Paris. Gauthier-Villars.

1971d Yu. V. Matijasevič. Diophantine representation of recursively enumerable predicates. In J. E. Fenstad, editor, *Proceedings of the Second Scandinavian Logic Symposium*, volume 63 of *Studies in Logic and the Foundations of Mathematics*, pages 171–177, Amsterdam. North-Holland.

Yu. V. Matiyasevich (continued)

1972a Diofantovy mnozhestva. *Uspekhi Matematicheskikh Nauk*, 27(5):185–222 (Russian). (Translated as Yu. V. Matiyasevich. Diophantine sets. *Russian Mathematical Surveys*, 27(5):124–164, 1972.)

1972b Arifmeticheskie predstavleniya perechislimykh mnozhestv s nebol'shim chislom kvantorov. *Zapiski Nauchnykh Seminarov Leningradskogo Otdeleniya Matematicheskogo Instituta im. V. A. Steklova AN SSSR (LOMI)*, 32:77–84 (Russian). (Translated as Yu. V. Matiyasevich. Arithmetical representations of enumerable sets with a small number of quantifiers. *Journal of Soviet Mathematics*, 6(4):410–416, 1976.)

1973 Yu. V. Matijasevič. On recursive unsolvability of Hilbert's Tenth Problem. In Patrick Suppes et al., editors, *Logic, Methodology and Philosophy of Science IV*, volume 74 of *Studies in Logic and the Foundations of Mathematics*, pages 89–110, Amsterdam. North Holland.

1974 Sushchestvovanie neèffektiviziruemykh otsenok v teorii èksponentsial'no diofantovykh uravneniĭ. *Zapiski Nauchnykh Seminarov Leningradskogo Otdeleniya Matematicheskogo Instituta im. V. A. Steklova AN SSSR (LOMI)*, 40:77–93 (Russian). (Translated as Yu. V. Matiyasevich. Existence of noneffectivizable estimates in the theory of exponential Diophantine equations. *Journal of Soviet Mathematics*, 8(3):299–311, 1977.)

1976 Novoe dokazatel'stvo teoremy ob èksponentsial'no diofantovom predstavlenii perechislimykh predikatov. *Zapiski Nauchnykh Seminarov Leningradskogo Otdeleniya Matematicheskogo Instituta im. V. A. Steklova AN SSSR (LOMI)*, 60:75–92 (Russian). (Translated as Yu. V. Matiyasevich. A new proof of the theorem on exponential Diophantine representation of enumerable sets. *Journal of Soviet Mathematics*, 14(5):1475–1486, 1980.)

1977a Prostye chisla perechislyayutsya polinomom ot 10 peremennykh. *Zapiski Nauchnykh Seminarov Leningradskogo Otdeleniya Matematicheskogo Instituta im. V. A. Steklova AN SSSR (LOMI)*, 68:62–82 (Russian). (Translated as Yu. V. Matijasevič. Primes are nonnegative values of a polynomial in 10 variables. *Journal of Soviet Mathematics*, 15(1):33–44, 1981.)

1977b Yu. V. Matijasevič. Some purely mathematical results inspired by mathematical logic. In Robert E. Butts and Jaakko Hintikka, editors, *Logic, Foundations of Mathematics, and Computability Theory*, volume 1 of *Proceedings of the Fifth International Congress of Logic, Methodology and Philosophy of Science*, pages 121–127, London, Ontario, Canada. D. Reidel Publishing Company, Dordrecht, Holland.

Yu. V. Matiyasevich (continued)

1977c Odin klass kriteriev prostoty, formuliruemykh v terminakh delimosti binomial'nykh coeffitsientov. *Zapiski Nauchnykh Seminarov Leningradskogo Otdeleniya Matematicheskogo Instituta im. V. A. Steklova AN SSSR (LOMI)*, 67:167–183 (Russian). (Translated as Yu. V. Matiyasevich. A class of primality criteria formulated in terms of the divisibility of binomial coefficients. *Journal of Soviet Mathematics*, 16(1):874–885, 1981.)

1979 Algorifmicheskaya nerazreshimost' èksponentsial'no diofantovykh uravneniĭ s tremya neizvestnymi. In A. A. Markov and V. I. Homič, editors, *Issledovaniya po teorii algorifmov i matematicheskoĭ logike*, volume 3, pages 69–78. Akademiya Nauk SSSR, Moscow. (Translated as Yu. V. Matiyasevich. Algorithmic unsolvability of exponential Diophantine equations in three unknowns. *Selecta Mathematica Sovietica*, 3(3):223–232, 1983/84.)

1984 Ob issledovaniyakh po nekotorym algorifmicheskim problemam algebry i teorii chisel. *Trudy Matematicheskogo Instituta im. V. A. Steklova AN SSSR*, 168:218–235 (Russian). (Translated as Yu. V. Matiyasevich. On investigations on some algorithmic problems in algebra and number theory. *Proceedings of Steklov Institute of Mathematics*, 168(3):227–252, 1986.)

1988 Diofantova slozhnost'. *Zapiski Nauchnykh Seminarov Leningradskogo Otdeleniya Matematicheskogo Instituta im. V. A. Steklova AN SSSR (LOMI)*, 174:122–131 (Russian). (Translated as Yu. V. Matiyasevich. Diophantine complexity. *Journal of Soviet Mathematics*, 55(2):1603–1610, 1991.)

1992 Yuri Matijasevich. My collaboration with Julia Robinson. *The Mathematical Intelligencer*, 14(4):38–45.

Yu. V. Matiyasevich and Julia Robinson

1974 Yuriĭ Matiyasevich and Dzhuliya Robinson. Dva universal'nykh trëkhkvantornykh predstavleniya perechislimykh mnozhestv. In B. A. Kušner and N. M. Nagornyĭ, editors, *Teoriya algorifmov i matematicheskaya logika*, pages 112–123. Vychislitel'nyĭ Tsentr, Akademiya Nauk SSSR, Moscow (Russian).

1975 Yuri Matijasevič and Julia Robinson. Reduction of an arbitrary Diophantine equation to one in 13 unknowns. *Acta Arithmetica*, 27:521–553.

Ernst W. Mayr and Albert R. Meyer

1981 The complexity of the finite containment problem for Petri nets. *Journal of the ACM*, 28(3):561–576.

Kenneth McAloon

1978 Completeness theorems, incompleteness theorems and models of arithmetic. *Transactions of the American Mathematical Society*, 239(5):253–277.

A. D. McGettrick

1977 Teaching mathematics by computer. *Computer Journal*, 20(3):263–268.

Z. A. Melzak

1960 A note on Hilbert's Tenth Problem. *Canadian Mathematical Bulletin*, 3(2):153–156.

1961 An informal arithmetical approach to computability and computation. *Canadian Mathematical Bulletin*, 4:279–293.

Friedhelm Meyer auf der Heide

1989 On genuinely time bounded computations. In B. Monien and R. Cori, editors, *STACS 89*, volume 349 of *Lecture Notes in Computer Science*, pages 1–16.

Žarko Mijajlović, Zoran Marković, and Kosta Došen

1986 *Hilbertovi problemi i logika*. Number 48 in Matematička Biblioteka. Zavod za Udžbenike i Nastavna Sredstva, Belgrade (Croation).

Charles F. Miller III

1973 Some connections between Hilbert's 10th Problem and the theory of groups. In W. W. Boone, F. B. Cannonito, and R. C. Lyndon, editors, *Word Problems: Decision Problems and the Burnside Problem in Group Theory*, volume 71 of *Studies in Logic and the Foundations of Mathematics*, pages 483–506. North-Holland, Amsterdam.

Marvin L. Minsky

1961 Recursive unsolvability of Post's problem of "tag" and other topics in the theory of Turing machines. *Annals of Mathematics, Second Series*, 74(3):437–455.

1967 *Computation: Finite and Infinite Machines*. Prentice-Hall, Englewood Cliffs, New Jersey.

G. E. Mints

1975 Teoriya dokazatel'stv (arifmetika i analiz). In R. V. Gamkrelidze, editor, *Algebra. Topologiya. Geometriya*, volume 13, pages 5–49. Akademiya Nauk SSSR. Vsesoyuznyĭ Institut Nauchnoĭ i Tekhnicheskoĭ Informatsii, Moscow (Russian). (Translated as G. E. Mints. Theory of proofs (arithmetic and analysis). *Journal of Soviet Mathematics*, 7(4):501–531, 1977.)

J. Mlček

1976 Twin prime problem in an arithmetic without induction. *Commentationes Mathematicae Universitatis Carolinae*, 17(3):543–555.

Joel Moses

1971 Algebraic simplification: A guide for the perplexed. *Communications of the ACM*, 14(8):527–537.

Nobuyoshi Motohashi

1984 A normal form theorem for first order formulas and its application to Gaifman's splitting theorem. *Journal of Symbolic Logic*, 49(4):1262–1267.

Albert A. Mullin

1974 Research problem 7. *Canadian Mathematical Bulletin*, 17(1):149.

A. G. Myasnikov and V. N. Remeslennikov

1986 Elementary group equivalence with the integral length function. *Illinois Journal of Mathematics*, 30(2):335–354.

J. Myhill

1953 Three contributions to recursive function theory. In *Actes du XIième Congrès International de Philosophie*, volume XIV, pages 50–59, Brussels.

A. Nerode

1963 A decision method for p-adic integral zeros of Diophantine equations. *Bulletin of the American Mathematical Society*, 69:513–517.

Jakob Nielsen

1924 Die Gruppe der dreidimensionalen Gittertransformationen. *Danske Videnskabernes Selskab Mathematisk-fysiske Meddelelser*, 5(12):1–29 (German).

Leszek Pacholski and Wiesław Szwast

1991 Asymptotic probabilities of existential second-order Gödel sentences. *Journal of Symbolic Logic*, 56(2):427–438.

Peter Pappas

1985 A Diophantine problem for Laurent polynomial rings. *Proceedings of the American Mathematical Society*, 93(4):713–718.

Jeff B. Paris and Constantine Dimitracopoulos

1982 Truth definitions for Δ_0 formulae. In *Logic and Algorithmic*, Monographie N° 30 de L'Enseignement Mathématique, pages 317–329. Université de Genève.

Yu. G. Penzin

1973 Razreshimost' teorii tselykh chisel so slozheniem, poryadkom i umnozheniem na proizvol'noe chislo. *Matematicheskie Zametki*, 13(5):667–675 (Russian). (Translated as Yu. G. Penzin. Solvability of the theory of integers with addition, order, and multiplication by an arbitrary number. *Mathematical Notes*, 13(5):401–405, 1973.)

1978 Nerazreshimye teorii kol'tsa nepreryvnykh funktsiĭ. *Algoritmicheskie voprosy algebraicheskikh sistem*, pages 142–147. Irkutsk (Russian).

1979 Problema bliznetsov v formal'noĭ arifmetike. *Matematicheskie Zametki*, 26(4):505–511 (Russian). (Translated as Yu. G. Penzin. Twins problem in formal arithmetic. *Mathematical Notes*, 26(3–4):743–746, 1979.)

Thanases Pheidas

1987a An undecidability result for power series rings of positive characteristic. *Proceedings of the American Mathematical Society*, 99(2):364–366.

1987b An undecidability result for power series rings of positive characteristic. II. *Proceedings of the American Mathematical Society*, 100(3):526–530.

1988 Hilbert's Tenth Problem for a class of rings of algebraic integers. *Proceedings of the American Mathematical Society*, 104(2):611–620.

1991 Hilbert's Tenth Problem for fields of rational functions over finite fields. *Inventiones Mathematicae*, 103(1):1–8.

Emil L. Post

1936 Finite combinatory processes—formulation 1. *Journal of Symbolic Logic*, 1(3):103–105.

1944 Recursively enumerable sets of positive integers and their decision problems. *Bulletin of the American Mathematical Society*, 50:284–316.

Hilary Putnam

1960 An unsolvable problem in number theory. *Journal of Symbolic Logic*, 25(3):220–232.

Michael O. Rabin

1957 Effective computability of winning strategies. In M. Dresher, A. W. Tucker, and P. Wolff, editors, *Contributions to the Theory of Games. Volume III*, number 39 in *Annals of Mathematics Studies*, pages 147–157. Princeton University Press, Princeton, New Jersey.

Constance Reid

1970 *Hilbert*. Springer-Verlag, New York.

1986 The autobiography of Julia Robinson. *The College Mathematics Journal*, 17(1):3–21.

John H. Reif and Harry R. Lewis

1986 Efficient symbolic analysis of programs. *Journal of Computer and System Sciences*, 32(3):280–314.

V. N. Remeslennikov

1979 Ob odnoĭ algoritmicheskoĭ zadache dlya nil'potentnykh grupp i kolets. *Sibirskiĭ Matematicheskiĭ Zhurnal*, 20(5):1077–1081 (Russian). (Translated as V. N. Remeslennikov. An algorithmic problem for nilpotent groups and rings. *Siberian Mathematical Journal*, 20(5):761–764, 1980.)

N. N. Repin

1983 Uravneniya s odnoĭ neizvestnoĭ v nil'potentnykh gruppakh. *Matematicheskie Zametki*, 34(2):201–206 (Russian). (Translated as N. N. Repin. Equations with one unknown in nilpotent groups. *Mathematical Notes*, 34(1–2):582–585, 1984.)

Daniel Richardson

1968 Some undecidable problems involving elementary functions of a real variable. *Journal of Symbolic Logic*, 33(4):514–520.

Abraham Robinson

1973 Nonstandard arithmetic and generic arithmetic. In Patrick Suppes, Leon Henkin, Athanase Joja, and Gr. C. Moisil, editors, *Logic, Methodology and Philosophy of Science IV*, volume 74 of *Studies in Logic and the Foundations of Mathematics*, pages 137–154, Amsterdam. North-Holland.

Julia Robinson

1952 Existential definability in arithmetic. *Transactions of the American Mathematical Society*, 72(3):437–449.

1960 J. B. Robinson. The undecidability of exponential Diophantine equations. *Notices of the American Mathematical Society*, 7(1):75.

1962 The undecidability of exponential Diophantine equations. In Ernest Nagel, Patrick Suppes, and Alfred Tarski, editors, *Logic, Methodology and Philosophy of Science: Proceedings of the 1960 International Congress*, pages 12–13, Stanford. Stanford University Press.

1969a Unsolvable Diophantine problems. *Proceedings of the American Mathematical Society*, 22(2):534–538.

1969b Diophantine decision problems. In W. J. LeVeque, editor, *Studies in Number Theory*, volume 6 of *Studies in Mathematics*, pages 76–116. Mathematical Association of America.

1971 Hilbert's Tenth Problem. In *1969 Number Theory Institute*, volume 20 of *Proceedings of Symposia in Pure Mathematics*, pages 191–194, Providence, Rhode Island. American Mathematical Society.

1973 Solving Diophantine equations. In Patrick Suppes, Leon Henkin, Athanase Joja, and Gr. C. Moisil, editors, *Logic, Methodology and Philosophy of Science IV*, volume 74 of *Studies in Logic and the Foundations of Mathematics*, pages 63–67, Amsterdam. North-Holland.

Raphael M. Robinson

1951 Arithmetical definitions in the ring of integers. *Proceedings of the American Mathematical Society*, 2(2):279–284.

1956 Arithmetical representation of recursively enumerable sets. *Journal of Symbolic Logic*, 21(2):162–186.

1972 Some representations of Diophantine sets. *Journal of Symbolic Logic*, 37(3):572–578.

J. M. Robson

1990 Strong time bounds: Non-computable bounds and a hierarchy theorem. *Theoretical Computer Science*, 73(3):313–317.

V. A. Roman'kov

1977 O nerazreshimosti problemy èndomorfnoĭ svodimosti v svobodnykh nil'potentnykh gruppakh i v svobodnykh kol'tsakh. *Algebra i Logika*, 16(4):457–471 (Russian). (Translated as V. A. Roman'kov. Unsolvability of the endomorphic reducibility problem in free nilpotent groups and in free rings. *Algebra and Logic*, 16(4):310–320, 1978.)

1979 Ob universal'noĭ teorii nil'potentnykh grupp. *Matematicheskie Zametki*, 25(4):487–495 (Russian). (Translated as V. A. Roman'kov. Universal theory of nilpotent groups. *Mathematical Notes*, 25(3–4):253–258, 1979.)

Louis E. Rosier

1986 A note on Presburger arithmetic with array segments, permutation and equality. *Information Processing Letters*, 22(1):33–35.

B. V. Rozenblat

1985 Diofantovy teorii svobodnykh inversnykh polugrupp. *Sibirskiĭ Matematicheskiĭ Zhurnal*, 26(6):101–107 (Russian). (Translated as B. V. Rozenblat. Diophantine theories of free inverse semigroups. *Siberian Mathematical Journal*, 26(6):860–865, 1986.)

Robert S. Rumely

1986 Arithmetic over the ring of all algebraic integers. *Journal für die Reine und Angewandte Mathematik*, 368(5):127–133.

Keijo Ruohonen

1972 Hilbertin kymmenes probleema. *Arkhimedes*, no. 1–2:71–100 (Finnish).

1980 Hilberts tionde problem. *Nordisk Matematisk Tidskrift*, 28(4):145–154 (Swedish).

Arto Salomaa

1985 *Computation and Automata*, volume 25 of *Encyclopedia of Mathematics and Its Applications*. Cambridge University Press, Cambridge.

M. V. Sapir

1989 Minimal'noe mnogoobrazie assotsiativnykh algebr s nerazreshimoĭ problemoĭ ravenstva. *Matematicheskiĭ Sbornik*, 180(12):1691–1708 (Russian). (Translated as M. V. Sapir. A minimal variety of associative algebras with unsolvable word problem. *Mathematics of the USSR. Sbornik*, 68(2):567–584, 1991.)

M. V. Sapir and O. G. Kharlampovich

1989 Problema ravenstva v mnogoobraziyakh assotsiativnykh algebr i algebr Li. *Izvestiya Vysshikh Uchebnykh Zavedeniĭ Matematika*, 6(325):76–84 (Russian). (Translated as M. V. Sapir and O. G. Kharlampovich. Equality problem in varieties of associative algebras and Lie algebras. *Soviet Mathematics* (*Iz. VUZ*), 33(6):77–86, 1989.)

Bruno Scarpellini

1963 Zwei unentscheidbare Probleme der Analysis. *Zeitschrift für Mathematische Logik und Grundlagen der Mathematik*, 9(4):265–289 (German).

Lowell Schoenfeld

1976 Sharper bounds for the Chebyshev functions $\theta(x)$ and $\psi(x)$. II. *Mathematics of Computation*, 30(134):337–360. Corrigendum: Ibid., 30(136):900.

Wolfgang Schönfeld

1979 An undecidablility result for relation algebras. *Journal of Symbolic Logic*, 44(1):111–115.

Daniel Shanks

1972 Five number-theoretic algorithms. In R. S. D. Thomas and H. C. Williams, editors, *Proceedings of the Second Manitoba Conference on Numerical Mathematics*, volume VII of *Congressus Numerantium*, pages 51–70, University of Manitoba, Winnipeg, Manitoba, 5–7 October. Utilitas Mathematica Publishing. Winnipeg, Manitoba.

Harold N. Shapiro and Alexandra Shlapentokh

1989 Diophantine relationship between algebraic number fields. *Communications on Pure and Applied Mathematics*, 42(8):1113–1122.

J. C. Shepherdson and H. E. Sturgis

1963 Computability of recursive functions. *Journal of the ACM*, 10(2):217–255.

D. V. Shiryaev

1989 Nerazreshimost' nekotorykh algoritmicheskikh problem dlya nevetvyashchikhsya program. *Kibernetika*, no. 1:63–66 (Russian).

Alexandra Shlapentokh

1989 Extensions of Hilbert's Tenth Problem to some algebraic number fields. *Communications on Pure and Applied Mathematics*, 42(7):939–962.

1990 Diophantine definitions for some polynomial rings. *Communications on Pure and Applied Mathematics*, 43(8):1055–1066.

Carl Ludwig Siegel

1972 Zur Theorie der quadratischen Formen. *Nachrichten der Akademie der Wissenschaften in Göttingen. II. Mathematisch-Physikalische Klasse*, no. 3:21–46 (German).

J. Siekmann and P. Szabó

1989 The undecidability of the D_A-unification problem. *Journal of Symbolic Logic*, 54(2):402–414.

Michael F. Singer

1978 The model theory of ordered differential fields. *Journal of Symbolic Logic*, 43(1):82–91.

David Singmaster

1974 Notes on binomial coefficients. I, II, III. *The Journal of the London Mathematical Society (Second Series)*, 8(3):545–548, 549–554, 555–560.

Th. Skolem

1934 Über die Nicht-charakterisierbarkeit der Zahlenreihe mittels endlich oder abzählbar unendlich vieler Aussagen mit ausschliesslich Zahlenvariablen. *Fundamenta Mathematicae*, 23:150–161 (German).

Craig Smoryński

1977 C. Smorynski. A note on the number of zeros of polynomials and exponential polynomials. *Journal of Symbolic Logic*, 42(1):99–106.

1987 Diophantine encoding. Preprint 455, University of Utrecht, Department of Mathematics, February. Included in Smoryński [1991].

1991 *Logical number theory I: An Introduction*. Springer-Verlag, Berlin.

R. Statman

1981 On the existence of closed terms in the typed λ-calculus. II: Transformations of unification problems. *Theoretical Computer Science*, 15(3):329–338.

È. D. Stotskiĭ

1971a Formal'nye grammatiki i ogranicheniya na vyvod. *Problemy Peredachi Informatsii*, 7(1):87–101.

1971b Upravlenie vyvodom v formal'nykh grammatikakh. *Problemy Peredachi Informatsii*, 7(3):87–102.

1973 Beskontekstnye grammatiki s ogranichennoĭ primenimost'yu pravil vyvoda. In S. K. Shaumyan, editor, *Matematicheskaya lingvistika*, pages 114–129. Vsesoyuznyĭ Institut Nauchnoĭ i Tekhnicheskoĭ Informatsii, Moscow (Russian).

Bernd Sturmfels

1987 On the decidability of Diophantine problems in combinatorial geometry. *Bulletin of the American Mathematical Society* (*New Series*), 17(1):121–124.

Zhi-Wei Sun

1992 A new relation-combining theorem and its applications. *Zeitschrift für Mathematische Logik und Grundlagen der Mathematik*, 38(3):209–212.

Héctor J. Sussman

1971 *Hilbert's Tenth Problem*, volume 9 of *Revista Colombiana de Matemáticas, Monografías Matemáticas*. Sociedad Colombiana de Matemáticas, Bogotá.

Shuichi Takahashi

1974 *Méthodes Logiques en Géométrie Diophantienne*. Université de Montréal, Montreal (French).

S. P. Tarasov and L. G. Khachiyan

1980 Granitsy resheniĭ i algoritmicheskoĭ slozhnosti sistem vypuklykh diofantovykh neravenstv. *Doklady Akademii Nauk SSSR*, 255(2):296–300 (Russian). (Translated as S. P. Tarasov and L. G. Hačijan. Bounds of solutions and algorithmic complexity of systems of convex Diophantine inequalities. *Soviet Mathematics. Doklady*, 22(3):700–704, 1980.)

Alfred Tarski

1951 *A Decision Method for Elementary Algebra and Geometry.* University of California Press, Berkeley and Los Angeles.

Véronique Terrier

1990 Décidabilité de la théorie existentielle de N structuré par l'ordre naturel, la divisibilité, les prédicats puissances et les fonctions puissances. *Comptes Rendus de l'Académie des Sciences. Série I. Mathématique*, 311(12):749–752 (French).

Axel Thue

1909 Über Annäherungswerte algebraischer Zahlen. *Journal für die Reine und Angewandte Mathematik*, 135:284–305 (German).

Erik Tiden and Stefan Arnborg

1987 Unification problems with one-sided distributivity. *Journal of Symbolic Computation*, 3(1–2):183–202.

Shih-Ping Tung

1988 Definability on formulas with single quantifier. *Zeitschrift für Mathematische Logik und Grundlagen der Mathematik*, 34(2):105–108.

Paavo Turakainen

1978 A note on noncontext-free rational stochastic languages. *Information and Control*, 39(2):225–226.

1981 On some bounded semiAFLs and AFLs. *Information Sciences*, 23(1):31–48.

A. M. Turing

1936 On computable numbers, with an application to the Entscheidungsproblem. *Proceedings of the London Mathematical Society. Second Series*, 42:230–265.

1937 On computable numbers, with an application to the Entscheidungsproblem. A correction. *Proceedings of the London Mathematical Society. Second Series*, 43:544–546.

1939 Systems of logic based on ordinals. *Proceedings of the London Mathematical Society. Second Series*, 45:161–228.

V. A. Uspenskiĭ and A. L. Semënov

1987 *Teoriya algoritmov: osnovnye otkrytiya i prilozheniya.* Nauka, Moscow (Russian).

M. K. Valiev

1975 On polynomial reducibility of word problem under embedding of recursively presented groups in finitely presented groups. In J. Bečvář, editor, *Mathematical Foundations of Computer Science 1975*, volume 32 of *Lecture Notes in Computer Science*, pages 432–438. Springer-Verlag, September.

P. van Emde Boas

1983 Dominos are forever. In L. Priese, editor, *Report on the 1st GTI-workshop*, number 13, pages 75–95, Reihe Theoretische Informatik, Universität-Gesamthochschule Paderborn.

Yu. M. Vazhenin

1987 Algoritmicheskie problemy i ierarkhii yazykov pervogo poryadka. *Algebra i Logika*, 26(4):419–434 (Russian). (Translated as Yu. M. Vazhenin. Algorithmic problems and hierarchies of first-order languages. *Algebra and Logic*, 26(4):241–252, 1988.)

1988 Kriticheskie teorii. *Sibirskiĭ Matematicheskiĭ Zhurnal*, 29(1):23–31 (Russian). (Translated as Yu. M. Vazhenin. Critical theories. *Siberian Mathematical Journal*, 29(1):17–23, 1988.)

1989 Kriticheskie teorii nekotorykh klassov neassotsiativnykh kolets. *Algebra i Logika*, 28(4):393–401 (Russian). (Translated as Yu. M. Vazhenin. Critical theories of certain nonassociative rings. *Algebra and Logic*, 28(4):255–261, 1990.)

Ramarathnam Venkatesan and Sivaramakrishnan Rajagopalan

1992 Average case intractability of matrix and Diophantine problems. In *Proceedings of the Twenty-Fourth Annual ACM Symposium on the Theory of Computing*, pages 632–642, Victoria, British Columbia, Canada, 4–6 May.

A. K. Vinogradov and N. K. Kosovskiĭ

1975 Ierarkhiya diofantovykh predstavleniĭ primitivno rekursivnykh predikatov. *Vychislitel'naya tekhnika i voprosy kibernetiki*, no. 12:99–107. Lenigradskiĭ Gosudarstvennyĭ Universitet, Leningrad (Russian).

H. Wada

1975 Polynomial representations of prime numbers. *Sûgaku*, 27(2):160–161 (Japanese).

Paul S. Wang

1974 The undecidability of the existence of zeros of real elementary functions. *Journal of the ACM*, 21(4):586–589.

Alex Wilkie

1975 On models of arithmetic—answers to two problems raised by H. Gaifman. *Journal of Symbolic Logic*, 40(1):41–47.

1980 A. J. Wilkie. Applications of complexity theory to Σ_0-definability problems in arithmetic. In L. Pacholski, J. Wierzejewski, and A. J. Wilkie, editors, *Model Theory of Algebra and Arithmetic*, volume 834 of *Lecture Notes in Mathematics*, pages 363–369. Springer-Verlag.

George Wilmers

1985 Bounded existential induction. *Journal of Symbolic Logic*, 50(1):72–90.

J. Wolstenholme

1862 On certain properties of prime numbers. *Quarterly Journal of Pure and Applied Mathematics*, 5:35–39.

S. Yukna

1982 Arifmeticheskie predstavleniya klassov mashinnoĭ slozhnosti. *Matematicheskaya logika i eë primeneniya*, no. 2:92–107. Institut Matematiki i Kibernetiki Akademii Nauk Litovskoĭ SSR, Vil'nyus (Russian).

1983 Ob arifmetizatsii vychisleniĭ. *Matematicheskaya logika i eë primeneniya*, no. 3:117–125. Institut Matematiki i Kibernetiki Akademii Nauk Litovskoĭ SSR, Vil'nyus (Russian).

D. A. Zakharov

1970 Rekursivnye funktsii. Publications of the Department of Algebra and Mathematical Logic of the Novosibirsk State University 10, Novosibirsk. Gosudarstvenny/vi universitet., Novosibirsk.

1986 Diofantovost' rekursivno perechislimykh mnozhestv i predikatov (Russian). Appendix in A. I. Mal'tsev, 1986. *Algoritmy i rekursivnye funktsii*. Nauka, Moscow, second edition.

List of Notation

A, 19
α, 19
β, 19
E, 20
E, 169
H, 169
γ, 35
ϕ, 38
$\pi(n)$, 118
ψ, 118
Θ, 110
Ξ, 20
ζ, 117

ADD, 81
Add, 52
and, 74
APPEND, 81
arem, 12

B_n, 51
Bernoulli, 51
BIN, 97

Cantor, 41
Cantor_n, 41
Card, 129
Code, 44
Concat, 45
COPY, 81

D, 162
DEC, 80
DECODE, 82
deg, 112
DELETE, 80
div, 13
do, 74
DOUBLE, 97

ECard, 149
ECode, 60
Elem, 45
Elema, 41
Elemb, 41
$\text{Elem}_{n,m}$, 41
Eq, 47
EQUAL, 82
Equal, 47
erank, 156
erankiter, 156
ERASE, 97
Even, 11
explog, 120

F, 34
$\mathcal{F}_0$, 168
$\mathcal{F}_1$, 169
$\mathcal{F}_2$, 174
$\mathcal{F}_3$, 177
FIND, 79
Format, 59

$\mathbb{G}$, 139
G, 117
gcd, 13
GElem, 43
grem, 144

$\mathfrak{H}_0$, 66
$\mathfrak{H}_1$, 66

if, 74
INC, 80
iter, 156
itererank, 156

JUMP, 79

LAST, 79
lcm, 13
LEFT, 76
Less, 52

MARK, 80
Min, 53
MULT, 81
Mult, 52

$\mathbb{N}$, 139
NEVERSTOP, 78
NEW, 80
NEXT, 82
NOTEQUAL, 82
NotGreater, 47
NOTGREATER, 82
NP, 162

od, 74
Odd, 11
oper, 154
operord, 155
operordrank, 155
operrank, 155
ord, 153
ordoper, 155
ordrank, 153
ordrankoper, 155
Ortnorm, 50

P, 116
$\mathbf{P}_n$, 122
PNotGreater, 47
PNotGreater_p, 133
Prime, 46
Primenth, 52
PSmall, 47
PSmall_p, 134

$\mathbb{Q}$, 39

$\mathbb{R}$, 165
rank, 153
rankoper, 155
rankord, 153
rankordoper, 155
READ, 77
READNOT, 79
Real, 123
rem, 12
Repeat, 48
RESTORE, 81
RIGHT, 76

SCod, 61
SCode, 61
Small, 47
Solution, 63
STAR, 79
STOP, 78
Sum, 52

then, 74
THEREIS, 80
THEREWAS, 81

U, 57
U_0, 65
U_1, 58
U_n, 58
unit, 110

VACANT, 79

while, 74
WRITE, 77

$\mathbb{Z}$, 139

Name Index

Adamowicz, Zofia, 196, 221
Adleman, Leonard, 164, 221, 240
Adler, Andrew, 152, 179, 196, 216, 219, 221
Aĭvazyan, S. V., 196, 221
Amice, Yvette, 207, 221
Andrews, George E., 205, 221
Anick, David J., 196, 222
Arnborg, Stefan, 197, 253
Azra, Jean-Pierre, xx, 222

Baker, A., 222
Barton, D., 179, 222
Bauer, Friedrich L., 56, 222
Baur, Walter, 196, 222
Baxa, Christoph, 38, 222
Baxter, Lewis D., 196, 222
Beck, H., 197, 223
Bell, J. L., 223
Bernoulli, Jakob, 51
Bicknell-Johnson, Marjorie, 232
Blum, Lenore, 223
Blum, Manuel, 218, 223
Bokut', L. A., 196, 223
Bollman, Dorothy, 209, 223
Börger, E., xx, 223
Britton, J. L., 16, 223
Broy, Manfred, 197, 223

Calude, Cristian, 196, 224
Cantone, D., 196, 224
Cantor, Georg, 70
Carstens, Hans Georg, 197, 224
Caviness, B. F., 179, 224
Cegielski, Patrick, 224
Chaitin, Gregory J., 197, 224
Chan, Tat-hung, 197, 224
Chebyshev, P. L., 118
Chowla, S., 224
Chudnovsky, Gregory V., 38–39, 215, 225
Church, Alonzo, 99, 225
Clarke, Lori A., 197, 225
Cohen, Jacques, 197, 225

Conway, J. H., 207, 225
Cutello, V., 196, 224

Dańko, Wiktor, 197, 225
Davis, Martin, xx, 16, 17, 37–39, 54, 68–70, 99–101, 126, 127, 152, 197, 208, 209, 215, 216, 218, 225–227
Denef, J., 151, 152, 179, 216, 227
Dickson, Leonard Eugene, 205, 227
Dimitracopoulos, Constantine, 197, 229, 235, 246
Diophantus, 2, 16, 146, 227
Došen, Kosta, xx, 244
Dowling, Michael L., 228
Durnev, V. G., 196, 197, 228
Dyson, Verena H.
see Huber-Dyson, Verena

Euclid, 95
Euler, Leonard, 118, 199, 200

Farmer, William M., 228
Fenstad, Jens Erik, xx, 38, 228
Fermat, Pierre, 34, 38
Fibonacci, Leonardo, 38
Fitch, J. P., 179, 222, 228
Friedrichsdorf, Ulf, 197, 228

Gagliardi, G., 228
Gaifman, Haim, 197, 229
Garey, Michael R., 229
Geimanis, Dainis, 229
Germano, Giorgio, 197, 229
Gödel, Kurt, 53, 101, 126, 229
Goldbach, Christian, 117
Goldfarb, Warren D., 197, 229
Goodstein, R. L., 197, 230
Gorskiĭ, I. L., 197, 230
Guaspari, D., 197, 230
Gurari, Eitan M., 197, 230

Hack, Michel, 196, 197, 230
Hájek, Petr, 230

Harel, David, 197, 230
Hasenjaeger, G., 231
Hatcher, William S., 197, 231
Havel, Ivan, 231
Havránek, Tomáš, 230
Heath, Thomas L., 231
Heering, Jan, 197, 231
Hermes, Hans, xx, 231
Herrman, Oskar, 37, 231
Hilbert, David, xix–xxi, 1, 2, 4, 5, 7, 15, 16, 34, 38, 53, 54, 66, 69–71, 92–95, 97, 99–101, 116, 117, 122, 126, 127, 129, 138, 139, 146, 149, 151, 152, 162, 168, 169, 174, 179, 192, 196, 197, 207, 212, 232
Hirose, Ken, xx, 126, 215, 232
Hirschfeld, Joram, 197, 232
Hodgson, Bernard R., 164, 197, 231, 232, 236
Hoggatt, Jr., V. E., 232
Howell, Rodney R., 197, 233
Huber-Dyson, Verena, 196, 228, 233
Huet, Gérard, 197, 233
Huynh, Dung T., 197, 233

Ibarra, Oscar H., 197, 219, 230, 233
Iida, Shigeaki, 126, 215, 232

Jeroslow, R. G., 197, 233
Johnson, David S., 229
Jones, A. J., 207, 225
Jones, James P., xx, 55, 56, 69, 70, 101, 163, 195–197, 212, 215, 217–219, 228, 233–235
Joseph, Deborah, 197, 235
Jutila, Matti, 127

Kaplansky, Irving, xx, 235
Karhumäki, Juhani, 235
Kasami, Tadao, 197, 235
Kaye, Richard, 197, 235
Kent, Clement F., 164, 232, 236
Khachiyan, L. G., 197, 252
Kharlampovich, O. G., 196, 236, 250
Kim, K. H., 152, 236
Kiss, Péter, 236
Kleĭman, Yu. G., 196, 236
Kochen, Simon, 38
Kolmogorov, A. N., 99, 236
Koppel, Moshe, 236
Korablëva, N. B., 197, 228
Kosovskiĭ, N. K., 38, 164, 197, 208, 212, 237, 254
Kreisel, G., 68, 127, 237
Kryauchyukas, V. Yu., 56, 237
Kucherov, G. A., 197, 237
Kummer, E. E., 47, 52–54, 201, 238

Lagrange, J. L., 238
Lambek, Joachim, 101, 238
Laplaza, Miguel, 209, 223
Laski, Janusz, 197, 238
Lee, R. D., 197, 230
Leininger, Brian S., 197, 219, 233
Lenstra, Jr., H. W., 238
Levitz, Hilbert, 163, 164, 234, 238
Lew, John S., 209, 238
Lewis, Harry R., 197, 247
Lipshitz, Leonard, 151, 152, 179, 197, 216, 227, 239
Livesey, M., 197, 239
Lyndon, Roger C., 196, 239

Machover, M., 223
Macintyre, A., 197, 239
Mal'tsev, A. I., 100, 239
Maňas, Miroslav, 197
Manders, Kenneth L., 164, 221, 239, 240
Manevitz, Larry Michael, 197, 240
Manin, Yu. I., xx, 240
Margenstern, M., xx, 240
Marković, Zoran, xx, 244
Martin-Löf, Per, 179, 240
Mart'yanov, V. I., 240
Maslov, S. Yu., 100, 240
Máté, Atilla, 197, 241

Matiyasevich, Yu. V., xx, 17, 37, 38, 54, 55, 69, 101, 126, 127, 152, 162, 163, 208, 209, 212, 215–218, 226, 234, 241–243
Mayr, Ernst W., 197, 243
Maňas, Miroslav, 239
McAloon, Kenneth, 197, 243
McGettrick, A. D., 179, 244
Melzak, Z. A., 101, 244
Meyer auf der Heide, Friedhelm, 197, 244
Meyer, Albert R., 197, 243
Mijajlović, Žarko, xx, 244
Miller III, Charles F., 196, 244
Minsky, Marvin L., 101, 244
Mints, G. E., 197, 244
Mlček, J., 197, 245
Moses, Joel, 179, 245
Motohashi, Nobuyoshi, 197, 245
Mullin, Albert A., 245
Myasnikov, A. G., 196, 245
Myhill, J., 16, 245

Nerode, A., 245
Nielsen, Jakob, 211, 245

Oppen, Derek C., 197, 233

Pacholski, Leszek, 197, 245
Panchishkin, A. A., xx, 240
Pappas, Peter, 179, 246
Paris, Jeff B., 197, 235, 246
Păun, Gheorghe, 196, 224
Penzin, Yu. G., 179, 197, 246
Pepper, Peter, 197, 223
Petri, C., 196
Pheidas, Thanases, 151, 152, 246
Pnueli, Amir, 197, 230
Policrito, A., 196, 224
Post, Emil L., 99, 128, 246
Putnam, Hilary, 16, 37, 54, 68, 100, 126, 152, 209, 215, 226, 227, 247

Rabin, Michael O., 195–196, 247
Rajagopalan, Sivaramakrishnan, 164, 254
Reid, Constance, 16, 151, 247
Reif, John H., 197, 247
Remeslennikov, V. N., 196, 245, 247
Repin, N. N., 196, 247
Richardson, Daniel, 179, 247
Richardson, Debra J., 197, 225
Riemann, Georg F. B., 117–119, 121, 122, 126, 127
Robinson, Abraham, 197, 247
Robinson, Julia, 16, 17, 37, 38, 54, 68, 69, 100, 126–128, 163, 197, 207–209, 215, 217, 218, 226, 227, 243, 248
Robinson, Raphael M., 152, 162, 164, 216, 218, 248
Robson, J. M., 248
Roman'kov, V. A., 196, 249
Rosenberg, Arnold L., 209, 238
Rosier, Louis E., 197, 233, 249
Roush, F. W., 152, 236
Rozenblat, B. V., 196, 249
Rumely, Robert S., 151, 249
Ruohonen, Keijo, xx, 38, 249

Salomaa, Arto, xx, 249
Sapir, M. V., 196, 249, 250
Sato, Daihachiro, 55, 235
Scarpellini, Bruno, 179, 250
Schinzel, Andrzej, 127
Schoenfeld, Lowell, 127, 250
Schönfeld, Wolfgang, 250
Schupp, Paul E., 196, 239
Schütte, Kurt, 38
Semënov, A. L., 99, 253
Shanks, Daniel, 37, 250
Shapiro, Harold N., 151, 250
Shepherdson, John C., 101, 197, 228, 250
Shiryaev, D. V., 197, 219, 250
Shlapentokh, Alexandra, 151, 152, 250, 251
Shub, Mike, 223
Siegel, Carl Ludwig, 17, 251
Siekmann, J., 197, 239, 251
Simmons, H., 197, 239

Singer, Michael F., 179, 197, 251
Singmaster, David, 251
Skolem, Thoralf, 16, 251
Smale, Steve, 223
Smoryński, Craig, xx, 152, 216, 251
Statman, R., 197, 252
Staudt, Karl G. C. von, 210
Stavi, Jonathan, 197, 230
Stotskiĭ, È. D., 197, 252
Sturgis, H. E., 101, 250
Sturm, Jacque C. F., 168, 207
Sturmfels, Bernd, 252
Sun, Zhi-Wei, 205, 252
Sussman, Héctor J., xx, 252
Szabó, P., 197, 239, 251
Szwast, Wiesław, 197, 245

Takahashi, Shuichi, 252
Tarasov, S. P., 197, 252
Tarski, Alfred, 16, 37, 168, 207, 253
Terrier, Véronique, 253
Thue, Axel, 131, 253
Tiden, Erik, 197, 253
Tokura, Nobuki, 197, 235
Tulipani, S., 228
Tung, Shih-Ping, 164, 253
Turakainen, Paavo, 197, 253
Turing, Alan M., 99, 126, 127, 165, 179, 253

Unvericht, E., 197, 239
Uspenskiĭ, V. A., 99, 236, 253

Valiev, M. K., 196, 254
van Emde Boas, Peter, 101, 254
Vazhenin, Yu. M., 196, 254
Venkatesan, Ramarathnam, 164, 254
Vinogradov, A. K., 164, 254

Wada, Hideo, 55, 235, 254
Wang, Paul S., 179, 254
Wheeler, William, 197, 232
Wiens, Douglas, 55, 235
Wilkie, Alex J., 163, 197, 234, 255
Wilmers, George, 197, 255
Wirsing, Martin, 197, 223
Wolstenholme, J., 53, 255

Yakubovich, A. M., 197, 230
Yen, Hsu-Chun, 197, 233
Young, Paul, 197, 235
Yukna, S., 164, 255

Zakharov, D. A., xx, 255

Subject Index

Alphabet of a Turing machine, 71

Base of positional code, 45

Cantor number of a tuple, 41
Cipher
 of a polynomial, 60
 of a positional code of a tuple, 45
Code
 Gödel code of a tuple, 43
 of a polynomial, 60
 of a set, 57
 of an equation, 64
  extended, 60
  without parameters, 65
 positional code of a tuple, 44
Concatenation, 44
Configuration, 85

Davis normal form, 99
Davis's hypothesis, 99
Decision problem, 1
 individual subproblem, 1
Degree of an equation with respect to
 a given unknown, 1
 all unknowns, 1, 7
 all variables, 7
Dimension of a set, 7
Diophantine game, 181
Diophantine relation, 9
Diophantine representation
 of a function, 10
 of a property, 9
 of a relation, 9
 of a set, 7
 singlefold, 132
Diophantine term, 10
Dirichlet's Principle, multiplicative form, 125

Equation
 Diophantine, 1
  parametric, 7
  trigonometric, 15
 exponential Diophantine, 33
  unary, 33
  unary with fixed base, 34
 Pell, 205
 universal, 57
Equivalent codes, 47
Euler's Identity, 118, 199
Exponential Diophantine representation
 of a function, 33
 of a property, 33
 of a relation, 33
 of a set, 33

Family of Diophantine equations, 6
Function
 Chebyshev's, 118
 Diophantine, 10
 probe, 155
 Turing computable, 97
Fundamental Theorem of Arithmetic, 118

Gaussian integers, 138
Goldbach's Conjecture, 117

Instruction
 of a chess machine, 186
 of a register machine, 98
 of a Turing machine, 72
Iteration level
 of a set, 156
  relative, 156
 of an equation, 156

Julia Robinson predicate, 37

Length of positional code, 45

Machine
 chess, 186
 Diophantine, 164
 Turing, 71
  nondeterministic, 164

Non-effective estimate, 131

Number of an equation, 123
Numbers
 Diophantine real, 165
 Fibonacci, 38
 natural, 5
 triangular, 205

Order of a set, 153
 relative, 153, 155

Parameter, 7
 code parameter, 57
 element parameter, 57
Prime Number Theorem, 118
Primes
 Fermat, 56
 Mersenne, 56
Property
 Diophantine, 9
 polynomial, 126

Rank of a set, 153
 exponential, 156
  relative, 156
 relative, 153, 155
Realization of a sequence of polynomials, 123
Register
 of a chess machine, 186
  input, 187
  output, 187
 of a register machine, 98
Relation of exponential growth, 35
Relation of reachability, 184
Relation of roughly exponential growth, 36
Relative number of operations, 155
Representation of a number, 75
 binary, 97
Representation of a tuple, 75
 binary, 97
 canonical, 75
Riemann Hypothesis, 117
Riemann's zeta function, 117

Set
 (A, C, X)-Diophantine, 139
 binary Turing semidecidable, 97
 decidable in the intuitive sense, 95
 Diophantine, 7
 of reachability, 184
 semidecidable in the intuitive sense, 95
  by a register machine, 98
 simple, 124
 tracing, 182
 Turing decidable, 92
 Turing semidecidable, 85
 univeral Diophantine, 57
State
 of a chess machine, 186
  final, 187
  initial, 186
 of Turing machine, 72
  final, 72
  initial, 72

Thesis
 Church's, 96
 Turing's, 99
Trivial solution of a homogeneous equation, 147
Tuple, 41
 characteristic, 49
 empty, 43

Universal bound
 on complexity, 154
 on number of operations, 154
 on order, 154
 on rank, 154
  on exponential rank, 156
Universal sequence of polynomials, 122
Unknown, 7

Zero of the zeta function
 non-trivial, 118
 trivial, 118